Parallel Science and Engineering Applications

The Charm++ Approach

Series in Computational Physics

Series Editors
Steven A. Gottlieb
Indiana University
Rubin H. Landau
Oregon State University

SERIES IN COMPUTATIONAL PHYSICS
Steven A. Gottlieb and Rubin H. Landau, Series Editors

Parallel Science and Engineering Applications

The Charm++ Approach

Edited by
Laxmikant V. Kale
Abhinav Bhatele

CRC Press
Taylor & Francis Group
Boca Raton   London   New York

CRC Press is an imprint of the
Taylor & Francis Group, an **informa** business

CRC Press
Taylor & Francis Group
6000 Broken Sound Parkway NW, Suite 300
Boca Raton, FL 33487-2742

© 2014 by Taylor & Francis Group, LLC
CRC Press is an imprint of Taylor & Francis Group, an Informa business

No claim to original U.S. Government works

Printed on acid-free paper
Version Date: 20130923

International Standard Book Number-13: 978-1-4665-0412-7 (Hardback)

Library of Congress Cataloging-in-Publication Data

Parallel science and engineering applications : the Charm++ approach / editors,
Laxmikant V. Kale, Abhinav Bhatele.
      pages cm -- (Series in computational physics)
   Includes bibliographical references and index.
   ISBN 978-1-4665-0412-7 (hardback)
   1. Engineering--Data processing. 2. Science--Data processing. 3. Parallel processing
(Electronic computers) 4. Charm++ (Computer program language) I. Kale, Laxmikant
V. II. Bhatele, Abhinav.

TA345.5.P37P37 2013
502.85'5133--dc23   2013034930

Visit the Taylor & Francis Web site at
http://www.taylorandfrancis.com

and the CRC Press Web site at
http://www.crcpress.com

To the generations of students and staff
at the Parallel Programming Laboratory and
our collaborators.

Contents

List of Figures

List of Tables

Series in Computational Physics

There can be little argument that computation has become an essential element in all areas of physics, be it via simulation, symbolic manipulations, data manipulations, equipment interfacing or something with which we are not yet familiar. Nevertheless, even though the style of teaching and organization of subjects being taught by physics departments has changed in recent times, the actual content of the courses has been slow to incorporate the newfound importance of computation. Yes, there are now speciality courses and many textbooks in computational physics, but that is not the same thing as incorporating computation into the very heart of a modern physics curriculum so that the physics being taught today more closely resembles the physics being done today. Not only will such integration provide valuable professional skills to students, but it will also help keep physics alive by permitting new areas to be studied and old problems to be solved.

This series is intended to provide undergraduate- and graduate-level textbooks for a modern physics curriculum in which computation is incorporated within the traditional subjects of physics, or in which there are new, multidisciplinary subjects in which physics and computation are combined as a "computational science." The level of presentation will allow for their use as primary or secondary textbooks for courses that wish to emphasize the importance of numerical methods and computational tools in science. They will offer essential foundational materials for students and instructors in the physical sciences as well as academic and industry professionals in physics, engineering, computer science, applied math and biology.

Titles in the series are targeted to specific disciplines that currently lack a textbook with a computational physics approach. Among these subject areas are: condensed matter physics, materials science, particle physics, astrophysics, mathematical methods of computational physics, quantum mechanics, plasma physics, fluid dynamics, statistical physics, optics, biophysics, electricity and magnetism, gravity, cosmology and high-performance computing in physics. We aim for a presentation that is concise and practical, often including solved problems and examples. The books are meant for teaching, although researchers may find them useful as well. In select cases, we have allowed more advanced, edited works to be included when they share the spirit of the series—to contribute to wider application of computational tools in the classroom as well as research settings.

Although the series editors had been all-too-willing to express the need for change in the physics curriculum, the actual idea for this series came from the series manager Luna Han of Taylor & Francis Publishers. We wish to thank her sincerely for that, as well as for encouragement and direction throughout the project.

Steven A. Gottlieb, Bloomington
Rubin H. Landau, Corvallis
Series Editors

Foreword

Laxmikant (Sanjay) V. Kale and Abhinav Bhatele have developed an eminently readable and comprehensive book that provides the very first in-depth introduction in book form to the research that Sanjay Kale has pioneered in the last 20 years with his research group at the University of Illinois. This work results from the successful interaction of two highly important topics of relevance to computational physics: parallel programming systems and highly scalable applications, in particular dynamic applications. The book provides a broad background on Charm++, the programming system developed by Kale and his group, as well as a number of highly relevant application studies that show the power and effectiveness of the Charm++ approach on a number of challenging problems in computational physics. As we move from the petascale era to exascale computing, it becomes increasingly clear that new paradigms are needed, that will increase our ability to handle adaptivity and dynamicism in applications and that go beyond the standard simple model provided by MPI. I expect Charm++ to play a role of even more importance in the future, and this book will significantly contribute to the wider distribution of its key concepts in the computational science community.

About two decades ago Sanjay Kale started the development of Charm++, a methodology for programming massively parallel machines. This was based on the observation that high productivity and efficiency can be attained if the distribution and coordination of computational work is considered at a higher level of abstraction. Such abstraction is independent of the number of computational cores and other idiosyncrasies of parallel machines' architectures, and allows expressing parallelism in terms natural to the problem being solved, instead of terms specific to the underlying machine being used. Although the performance and scalability of any parallel algorithm depends upon an even distribution of computational work and the time required synchronizing the distributed work, programmer productivity depends upon minimizing the amount of time required to implement such algorithms on the available computational hardware. Once the problem is appropriately expressed, the specific details of distributing, scheduling and coordinating the work on any particular computational resource can then be delegated to a suitable adaptive runtime system. Charm++ provides such a runtime system, with advanced mechanisms for optimized communication and measurement-based dynamic load balancing.

Charm++ and its main idea of an adaptive runtime system have significantly impacted the state of the art in parallel computing. Charm++ is now in widespread use on clusters, and it accounts for a significant fraction of CPU cycles used at multiple supercomputing centers, one of the very few academically developed software systems to do so. The original inspiration for developing Charm++ was molecular dynamics, and the NAMD software based on Charm++ has become one of the most significant application packages in the field. As the other chapters in the book show, Charm++ has been also applied to many other fields such as quantum chemistry, cosmology, weather forecasting, epidemiology, material science and others.

Charm++ was far ahead of its time, when we consider the state of the art in the early 1990s. Vector machines and Fortran were still very much entrenched, and MPI, the consensus model for using the MPPs of the day, was just about in the process of being defined. Proposing an approach based on C++ was novel, but assuming that one could develop a system that could automate many basic parallel programming tasks was daring. Many similar attempts had been made, and most of them have faded away in time. Charm++, however, prevailed and thrived. I think the reasons for success in the long term are closely related to the applications focus of Charm++, as is obvious in this book. Charm++ is not only a visionary approach developed by a computer scientist, but is also based on an understanding of how real applications are developed. Charm++ always was applied to real scientific computations and real HPC platforms, and that is one of the main reasons for its success and persistence in a research environment, where ideas come and go so quickly.

In the twenty years since the original development of Charm++, high performance computing has seen many dramatic developments. In the early 1990s computing technology had just made a dramatic transition to MPPs using commodity hardware and the MPI programming model. Essentially this model has remained the same until 2012, while at the same time increasing performance by a factor of one million from the gigaflop/s to the petaflop/s level. Today we are close to yet another transformation of the HPC field as GPUs and accelerators become integrated, while the amount of parallelism seems to be ever increasing.

In this context of a potential rapid transformation of the high performance computing field, the book by Kale and Bhatele arrives at exactly the right time. It succeeds perfectly and combines for the first time both Charm++ and significant application development in a single volume. It will provide a solid foundation for anyone who is considering using the most recent tools for developing applications for future exascale platforms. I highly recommend this timely book for scientists and engineers.

Horst Simon,
Lawrence Berkeley National Laboratory
and
University of California, Berkeley

Preface

Parallel computing, which started as a field from the early days of computing with systems such as ILLIAC IV, and became more widespread in the late 1980s with hypercubes, commodity clusters and symmetric multiprocessors, is poised for another large transition. With the use of multicore chips, all desktops, laptops and even mobile smartphones have become parallel computers. Large-end supercomputers have broken through the 50 petaflop/s peak performance barrier and in the middle, clusters are becoming ubiquitous within departments or in the cloud infrastructures. Looking to the future, the machines are likely to become more complex because of power considerations and reliability issues brought on by device scaling. At the same time, the applications are becoming more sophisticated: by simulating dynamic behaviors of physical systems with adaptive data structures, and incorporating multiple physics modules, for example.

Some of the ideas that are increasingly being discussed in the field have to do with increased support for asynchrony in the programming models, message-driven execution and macro-dataflow, automatic load balancing and efficient fault tolerance. It so happens that the Charm++ parallel programming system was developed to support such dynamic applications and provide the features mentioned above, and is ready to meet the challenges of the coming era in adaptive parallel computing. We think that it represents a technology that is maturing just as the need for it is being acutely apparent. To be sure, to scale to the next generation of machines, and to handle the next generation hardware quirks, research and development will be needed to enhance its adaptive runtime and to raise its level of abstraction. But the basic programming model and its underlying execution model it represents is a high-potential candidate in this era.

Charm++ is a production-quality system, with the ability to run on almost any parallel computer available and with a robust nightly build/test system. It was developed in the context of computational science and engineering (CSE) applications, with each abstraction motivated by and further honed by specific application needs. Therefore, we believe that this book is very timely and useful as it surveys a diverse collection of scalable CSE applications, most of which are used regularly on supercomputers by scientists to further their research. The book can be used by scientists, application developers and students to understand the Charm++ programming model and its family of mini-languages such as Adaptive MPI, and to scrutinize and ascertain its utility including for future applications and libraries. It is worth noting that Charm++ libraries can now be used in MPI applications with an interoperability model that makes it easy to incrementally start using Charm++.

The main theme of the book is the CSE applications developed using Charm++. However, given the above objectives, we have included three chapters in the beginning that introduce and review the Charm++ model, techniques used in designing Charm++ applications, and debugging and performance analysis tools included with Charm++.

The book begins with an introduction to the Charm++ programming model in the first chapter. Charm++ is a parallel programming system developed over the past fifteen years or so at the University of Illinois at Urbana-Champaign. It is aimed at enhancing productivity in parallel programming while enhancing scalable parallel performance. A guiding principle behind the design of Charm++ is to automate what the "system" can do best, while leaving to the programmers what they can do best. In particular, the programmer can specify *what* to do in parallel relatively easily, while the system can best decide which processors own which data units and execute which work units. This approach requires an *intelligent runtime system*, which Charm+ provides.

Thus, Charm++ employs the idea of "over-decomposition" or "processor virtualization" based on *migratable objects*. This idea leads to programs that automatically respect locality, in part because objects provide a natural encapsulation mechanism. At the same time, it empowers the runtime system to automate resource management. The combination of features in Charm++ and associated languages makes them suitable for the expression of parallelism for a range of architectures, from desktops to Petaflop/s scale parallel machines.

The second chapter discusses the design methodology for Charm++ programs, and illustrates the use of various features of Charm++ via a series of examples. Writing parallel programs using the Charm++ programming model is substantially different from the prevalent models such as MPI and OpenMP. An essential difference arises because Charm++ allows the decomposition to be independent of the number of processors and because of the asynchronous nature of the communication. This chapter discusses the design choices one makes when developing Charm++ applications.

The third chapter discusses tools for debugging and performance analysis of Charm++ programs. The tools are the CHARMDEBUG debugger and the PROJECTIONS performance analysis and visualization suite.

The focus of the book is on Charm++ applications and the next seven chapters (4-10) form the heart of the book, each devoted to one application, discussing the parallel design considerations and challenges. The book presents several parallel CSE codes written in the Charm++ model, along with the science and the numerical formulations underlying them and explains their parallelization strategies and parallel performance. These chapters demonstrate the versatility of Charm++ and its utility for a wide variety of applications, including molecular dynamics, cosmology, quantum chemistry, fracture simulations, agent-based simulations, and weather modeling. We hope that parallelization strategies used for specific application classes will give insights into developing similar CSE applications.

The molecular dynamics application NAMD is presented in the fourth chapter. NAMD (NAnoscale Molecular Dynamics) is a parallel molecular dynamics (MD) code designed for high performance simulation of large biomolecular systems. Typical NAMD simulations include all-atom models of proteins, lipids and/or nucleic acids as well as explicit solvent (water and ions) and range in size from 10,000 to 10,000,000 atoms.

NAMD employs the prioritized message-driven execution capabilities of the Charm++ parallel runtime system, along with dynamic load balancing, to attain excellent parallel scaling on both massively parallel supercomputers and commodity workstation clusters. NAMD is distributed free of charge as both source code and pre-compiled binaries by the Theoretical and Computational Biophysics Group of the University of Illinois Beckman Institute. The chapter on NAMD outlines its design and parallelization and its current performance on the largest supercomputers at different centers.

The fifth chapter discusses a quantum chemistry application called OPEN-ATOM. An accurate understanding of phenomena occurring at the quantum scale can be achieved by considering a model representing the electronic structure of the atoms involved. The CPAIMD method is one such algorithm, which has been widely used to study systems containing tens to thousands of atoms. To achieve a fine-grained parallelization of CPAIMD, computation in OPE-NATOM is divided into a large number of objects, enabling scaling to tens of thousands of processors. OPENATOM is a communication-intensive application with several phases and is an interesting case study of using the Charm++ programming model. Topology aware mapping of OPENATOM tasks is a widely applicable technique for performance optimizations which is highlighted in this chapter.

The sixth chapter presents the cosmology simulator, CHANGA. CHANGA is an N-body code that models gravity as well as smooth particle hydrodynamics (SPH). It leverages object-based virtualization and the data-driven style of computation of Charm++, to obtain adaptive overlap of communication and computation, as well as to perform automatic measurement-based load balancing. CHANGA advances the state-of-the-art in N-body simulations by allowing the programmer to achieve higher levels of resource utilization on large systems.

A related application for cosmological visualizations, Salsa, is discussed in the next chapter. Visualization in the context of cosmology involves a huge amount of data, possibly spread over multiple processors. This problem is solved by using a parallel visualizer written in Charm++. It uses the client-server model in Charm++ to interact with code running on a parallel computer. The visualizer allows the user to zoom in and out, pan and rotate the image. This results in update requests being sent from the visualizer code to the parallel programs. The chapter also describes a series of optimizations that have been carried out to this basic scheme.

The eighth chapter presents BRAMS, a parallel application for weather simulations. This chapter demonstrates the effectiveness of processor virtualization for dynamically balancing the load in BRAMS, a mesoscale weather forecasting model based on MPI. BRAMS uses the Charm++ infrastructure, with its over-decomposition and object-migration capabilities, to move subdomains across processors during program execution. Processor virtualization enables better overlap between computation and communication and improved cache efficiency. Furthermore, employing an appropriate load balancer results in better processor utilization while requiring minimal changes to the model's code.

BRAMS uses Adaptive MPI (AMPI) which facilitates running MPI programs as Charm++ programs, thereby providing the benefits of load balancing, fault tolerance and others to MPI applications. We use this opportunity to briefly introduce AMPI in this chapter.

Crack propagation and the associated application, Fractography3D, are discussed in the ninth chapter. Fractography3D is a dynamic 3D crack propagation simulation tool written using the FEM framework. FEM is a framework built on top of Charm++ for applications based on finite element methods. This chapter describes the design of the FEM framework briefly and factoring big problems between computer and computational scientists. Then, it discusses the basic fracture physics, and issues with partitioning meshes for very large machines (and ParMETIS).

The last chapter in the book presents an application that is different from the standard CSE applications widely used in the high performance computing community. EpiSimdemics, an agent-based simulation of people and places, is used to study the impact of agent behavior and public policy mitigation strategies over extremely large interaction networks. This involves the simulation of a co-evolving graph-based discrete dynamical system which is achieved by using a simple parallel discrete event simulation. The unpredictable nature of the graph in this application presents a challenging scenario for load balancing. The chapter discusses the implementation of the EpiSimdemics algorithm using Charm++ and presents some of the best performance numbers for this type of application.

The book is intended for a broad audience in academia and industry associated with the field of high performance computing. Application developers and users will find this book interesting as an introduction to Charm++ and to developing parallel applications in an asynchronous message-driven model. It will also be a useful reference for undergraduate and graduate courses in computer science and other engineering disciplines. Courses devoted to parallel programming and writing of parallel CSE applications will benefit from this book.

The book would not have been completed without the help of many people. The editors would like to thank all the reviewers who found time from their busy schedules to help improve the chapters—Adi Gundlapalli, Katherine Isaacs, Sohrab Ismail-Beigi, Atul Jain, Sriram Krishnamoorthy, Celso Mendes, Arun Prakash, Paul Ricker, Martin Schulz, Emad Tajkhorshid, Matthew Turk and Sathish Vadhiyar. Among the reviewers whom we asked to do some urgent reviews, two did their reviews in the same weeks as their first children were born! We thank them for their dedication. In addition to the PPL members involved in contributing chapters, we also want to thank others who helped with various editing tasks—JoAnne Geigner, Nikhil Jain, Akhil Langer, Phil Miller, Osman Sarood and many others.

We would also like to thank Horst Simon for writing a highly encouraging foreword, in a short time, in spite of his very busy schedule. We also thank the series editors, Steve Gottlieb and Rubin Landau, for their encouragement of this project. We are grateful to Luna Han for all the advice, assistance and guidance she provided, and to the production team of Taylor & Francis, in creating a well-crafted book.

The first editor (Kale) would like to thank his family, and especially his wife Lata, for the support and encouragement for the work in putting the book together and during years of research that the book reports on. The second editor (Bhatele) would like to thank his family for believing in him, appearing disappointed at his successes and saying, "you could've done better."

Laxmikant V. Kale, Urbana
Abhinav Bhatele, Livermore
Editors

About the Editors

Laxmikant V. ("Sanjay") Kale is the director of the Parallel Programming Laboratory and a professor of computer science at the University of Illinois at Urbana-Champaign. Dr. Kale has been working on various aspects of parallel computing, with a focus on enhancing performance and productivity via adaptive runtime systems, with research on dynamic load balancing, fault tolerance and power management. Underlying his work is the belief that only interdisciplinary research involving multiple CSE and other applications can bring back well-honed abstractions into computer science that will have a long-term impact on state of the art. His collaborations include the widely used Gordon-Bell award winning (Supercomputing 2002) biomolecular simulation program, NAMD and other collaborations on computational cosmology code (CHANGA), quantum chemistry (OPENATOM), agent-based simulation (EPISIMDEMICS), rocket simulation, space-time meshes and other unstructured mesh applications. He takes pride in his group's success in distributing and supporting software embodying his research ideas, including Charm++, Adaptive MPI and associated tools. His team won the HPC Challenge award at Supercomputing 2011, for its entry based on Charm++.

Dr. Kale received a B.Tech. degree in electronics engineering from Benares Hindu University, Varanasi, India, in 1977, and an M.E. degree in computer science from Indian Institute of Science in Bangalore, India, in 1979. He received a Ph.D. in computer science from the State University of New York, Stony Brook, in 1985. He worked as a scientist at the Tata Institute of Fundamental Research from 1979 to 1981. He joined the faculty of the University of Illinois at Urbana-Champaign as an assistant professor in 1985, where he is currently employed as a professor. Dr. Kale is a fellow of the IEEE and a recipient of the IEEE Sidney Fernbach award for 2012.

Abhinav Bhatele is a computer scientist at
the Center for Applied Scientific Computing at
Lawrence Livermore National Laboratory. His in-
terests lie in performance optimizations through
analysis, visualization and tuning and developing algorithms for high-end parallel systems. His
thesis was on topology aware task mapping and
distributed load balancing for parallel applications.

Dr. Bhatele received a B.Tech. degree in computer science and engineering from I.I.T. Kan-
pur, India, in May 2005, and M.S. and Ph.D. de-
grees in computer science from the University of Illinois at Urbana-Champaign
in 2007 and 2010, respectively. Dr. Bhatele was an ACM/IEEE-CS George
Michael Memorial HPC Fellow in 2009. He has received several awards for his
dissertation work including the David J. Kuck Outstanding MS Thesis Award
in 2009, a Distinguished Paper Award at Euro-Par 2009 and the David J.
Kuck Outstanding PhD Thesis Award in 2011. Recently, a paper that he co-
authored with LLNL and external collaborators was selected for a Best Paper
award at IPDPS in 2013.

Contributors

Ashwin M. Aji
Virginia Tech
Blacksburg, Virginia

Keith R. Bisset
Virginia Tech
Blacksburg, Virginia

Eric J. Bohm
University of Illinois at Urbana-
   Champaign
Urbana, Illinois

M. Scot Breitenfeld
University of Illinois at Urbana-
   Champaign
Urbana, Illinois

Philippe H. Geubelle
University of Illinois at Urbana-
   Champaign
Urbana, Illinois

Filippo Gioachin
Hewlett-Packard Laboratories
Singapore

Abhishek Gupta
University of Illinois at Urbana-
   Champaign
Urbana, Illinois

Pritish Jetley
University of Illinois at Urbana-
   Champaign
Urbana, Illinois

Tariq Kamal
Virginia Tech
Blacksburg, Virginia

Orion Sky Lawlor
University of Alaska
Fairbanks, Alaska

Chee Wai Lee
Texas A&M University
College Station, Texas

Jonathan Lifflander
University of Illinois at Urbana-
   Champaign
Urbana, Illinois

Madhav V. Marathe
Virginia Tech
Blacksburg, Virginia

Glenn J. Martyna
IBM Research
Yorktown Heights, New York

Chao Mei
Intel Corporation
Champaign, Illinois

Celso L. Mendes
University of Illinois at Urbana-
  Champaign
Urbana, Illinois

Jairo Panetta
Instituto Tecnologico de
  Aeronautica
São José dos Campos, Brazil

James C. Phillips
University of Illinois at Urbana-
  Champaign
Urbana, Illinois

Thomas R. Quinn
University of Washington
Seattle, Washington

Eduardo R. Rodrigues
IBM Research
São Paulo, Brazil

Klaus Schulten
University of Illinois at Urbana-
  Champaign
Urbana, Illinois

Yanhua Sun
University of Illinois at Urbana-
  Champaign
Urbana, Illinois

Ramprasad Venkataraman
University of Illinois at Urbana-
  Champaign
Urbana, Illinois

Jae-Seung Yeom
Virginia Tech
Blacksburg, Virginia

Gengbin Zheng
University of Illinois at Urbana-
  Champaign
Urbana, Illinois

Chapter 1

The Charm++ Programming Model

Laxmikant V. Kale

Department of Computer Science, University of Illinois at Urbana-Champaign

Gengbin Zheng

National Center for Supercomputing Applications, University of Illinois at Urbana-Champaign

Charm++ [126] is a C++ based parallel programming system developed at the University of Illinois. It has been designed and refined in the context of collaborative development of multiple science and engineering applications, as the later chapters in this book illustrate. The signature strength of Charm++ is its *adaptive runtime system*, which allows programmers to deal with increasingly complex supercomputers and sophisticated algorithms with dynamic and evolving behavior. Its basic innovation is the idea of *over-decomposition* (explained further in Section 1.2): the programmer decomposes the computation into objects rather than processors, and leaves the decision about which object lives on which processor to the runtime system. Specifically, some of the benefits of Charm++ to the programmer include:

- *Processor-independent programming:* The programmer decomposes the computation into logical units that are natural to the application, uncluttered by the notion of what data is found on which processor, and which computations happen on which processor.

- *Asynchronous programming model with message-driven execution:* Communication is expressed in a highly asynchronous manner, without opportunities for the program to block the processor awaiting a remote event. This model is supported by message-driven execution, where the processor-level scheduler selects only those computations for which all data dependencies have been satisfied.

- *Automatic communication/computation overlap:* Without any explicit programming effort, the Charm++ runtime ensures that processors are not held up for communication, and that the communication is spread more uniformly over time. This leads to better utilization of the communication network, and to programs that are highly tolerant of communication latencies.

- *Load Balancing:* The Charm++ runtime automatically balances load, even for applications where the load evolves dynamically as application progresses. It can handle both machine-induced and application-induced load imbalances.

- *Resilience:* Charm++ applications can be automatically checkpointed to disk, and restarted on a different number of processors, within memory limits. Further, Charm++ applications can be made automatically tolerant of node failures, automatically restarting based on in-memory checkpoints when the system detects a failure, on machines where the schedulers will not kill a job if one node dies.

The purpose of this chapter is to introduce the basic concepts in Charm++, and describe its capabilities and benefits for developing complex parallel applications using it. The next chapter illustrates the process of designing applications in Charm++ with choices and design decisions one must make along the way. A more thorough tutorial on how to design Charm++ applications can be found elsewhere. For example, for online resources, see `http://charm.cs.illinois.edu`.

1.1   Design Philosophy

To appreciate the features of Charm++, it is necessary to understand the main design principles that were used as guidelines when developing it.

The first of these principles has to do with the question: *what aspects of the parallel programming task should the "system" automate?* The design of Charm++ is guided by the idea of seeking an *optimal division of labor* between the programmer and the system, i.e, we should let the programmers do what they can do best, and automate those aspects that are tedious for the programmer but relatively easy (or at least feasible) for a system to automate. Some parallel programming systems are designed with the view that the system should simply provide minimal mechanisms, such as basic communication primitives, and get out of the way. This has the advantage that the application developers are least constrained. An alternative is presented by the ideal of a perfect parallelizing compiler: the programmers write (or better still, just bring their own dusty deck) sequential code, and the system auto-magically parallelizes it effectively for the target machine. The former approach is inadequate because it does not raise the level of abstractions, while the latter has been seen to be unrealizable, despite valiant efforts. *Seeking an optimal division of labor* between the application programmer and the system has led to foundational design features in Charm++.

The second design principle is to develop features only in an application-driven manner. This is to counter a common and natural tendency among computer scientists toward a platonic approach to design, which one could call *design in a vacuum*: features are developed because they appear beautiful to their developers, without serious consideration of their relevance to a broad set of applications. To avoid this, Charm++ evolved in the context of development of parallel science and engineering applications, and abstractions or features were added to it when the application use cases suggested them [114].

1.2 Object-Based Programming Model

It is important to note that although Charm++ is certainly a novel and distinct parallel programming model, different than prevailing models such as MPI, it is **not** a different "language"—code is still written in C++.[1] Programmers are only required to provide declarations of the methods that are meant to be invoked remotely, so that the system can generate code to pack and unpack their parameters automatically. Beyond that, one uses the C++ API provided to make calls into the Charm++ runtime system.

The basic innovation in Charm++ is that the computation is broken down by the programmer into a large number of objects, independent of the number of processors. These objects interact with each other through asynchronous method invocations (defined below). In such interactions and communications, it is the objects that are named explicitly and not the processors; the program is mostly free of the concept of a *processor*. This empowers the adaptive runtime system at the heart of Charm++ to place objects on processors as it pleases, and to change the placement during the execution of the program. This *separation of concerns between the application logic and resource management* is at the heart of many benefits that this programming model confers on application developers.

These *"migratable"* objects, which are the units of decomposition in the parallel program, are called *chares*[2] in Charm++. Of course, a Charm++

---

[1]Although there exist bindings for C and Fortran, we will focus on the C++ bindings in this chapter. Most of the applications in the book are written in C++. It is also, of course, possible to write most of the application in C or Fortran by using Charm++ to express all the parallel aspects, and calling sequential C and Fortran functions for the application-specific code.

[2]The "a" in chare (\'tʃɑr\) is pronounced like the "a" in father and the "e" is silent.

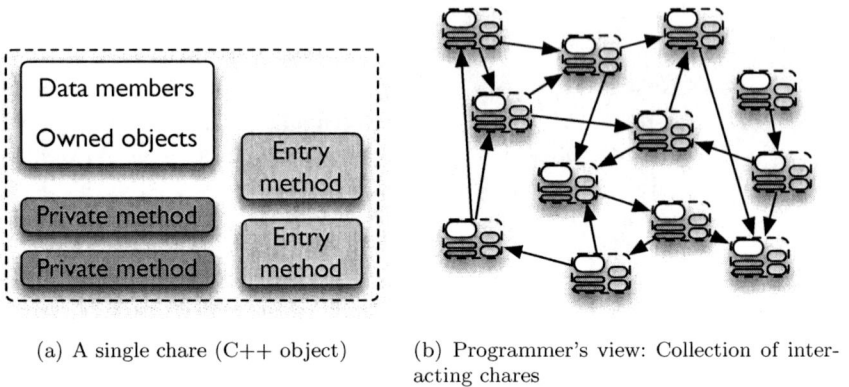

(a) A single chare (C++ object)   (b) Programmer's view: Collection of interacting chares

FIGURE 1.1: Chare: a message-driven object.

program may also include regular C++ objects—but the runtime system does not need to pay attention to them. Each such sequential regular C++ object is "owned" by a single chare (Figure 1.1(a)). So, they migrate with the chare if the runtime system decides to migrate the chare to another processor. The programmer's view of the overall computation is that of many such chares interacting with each other, as shown in Figure 1.1(b).

Let us examine a chare in isolation first, as shown in Figure 1.1(a): it is a C++ object comprising data elements and private or public methods. Its public methods are remotely invocable, and so are called "entry" methods. It is the existence of these entry methods that distinguishes a chare class from a plain C++ class. Chares can directly access their own data members, and cannot usually access data members of other chares. In that sense, a chare can be thought of as a processor as well. Since, typically, multiple chares live on a real processor, we can call them "virtual" processors. Consequently, we have called our approach the *"processor virtualization"* approach [122]; however, it is important to note that it is significantly different than (but related to) the relatively recent idea of *OS virtualization* made popular for the "cloud" infrastructure by VMWare and Xen systems.

Asynchronous method invocation: A running object may execute code that tells the runtime system to invoke an entry method on a (potentially) remote chare object with given parameters. The programmer understands that such method invocation is asynchronous: all that happens at the point where the call is made is that the parameters are packaged into a message, and the message is sent towards the chare object in question. It will execute at some undetermined point in the future. No return values are expected from an asynchronous method invocation. If needed, the called chare will send a method invocation to caller at some time in the future.

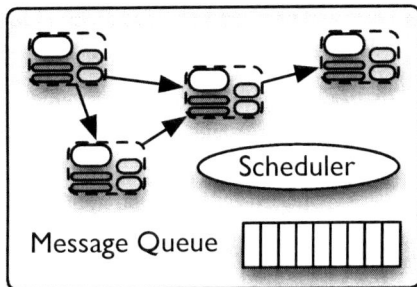

FIGURE 1.2: Message-driven scheduler.

The **execution model**, from the point of view of the programmer, is as follows: an ongoing computation consists of a collection of chare objects and a collection of entry method invocations that have been directed at these objects. The computation begins with construction of a designated *main chare*. The user code in the constructor of the main chare may initialize the *Read-only variables*. These should not be changed by user code afterwards. The runtime system makes a copy of such variables available on each processor. The constructor of the main chare typically contains user code that create chares and collections of chares (see below), and thus seeds the overall computation. On each processor, a message driven scheduler (Figure 1.2) in the runtime system selects one of the entry method invocations targeted at some object residing on its processor, unpacks the parameters for the invocation and executes the entry method on the identified object with the given parameters. In the baseline model, it lets the method invocation continue to completion (see Section 1.4 for exceptions), at which point control returns back to the runtime scheduler. Since the asynchronous method invocations can be thought of as messages, this aspect of the execution model is called *message-driven execution*.

One of the major benefits of message-driven execution is an automatic and adaptive overlap between communication and computation. There is no call in Charm++ that will block the *processor* waiting for some remote data. Instead, control passes to some chare that already has some data waiting for it, sent via the method invocation from a local or remote object. The time an object is waiting for some communication from its remote correspondent is thus naturally overlapped with computation for some other object that is ready to execute.

A chare normally just sits passively. Whenever a method invocation (typically initiated asynchronously by some other chare) arrives at a chare, it executes the method with the parameters sent in the invocation. This may result in creation of some new asynchronous method invocations for other chares (or even itself) that are handed over to the runtime system to deliver.

It changes its own state (i.e., values of its data member variables) as a result of the invocation. It then goes back to the passive state waiting for another method invocation.[3]

The code inside each entry method can carry out any computation it wishes using data completely local to its object. In addition, it can use data that has been declared as read-only.

From the application's point of view, a chare could be a piece of the domain (in "domain decomposition" methods used commonly in parallel CSE applications). It may also be a data structure, or a chunk of a matrix, or a work unit devoid of any persistent state. The programmer is responsible for deciding how big the chare should be, viz. the *grainsize* of the chare. More on that in the next chapter.

One can create a singleton chare instance dynamically, and the system will decide on which processor to anchor it. All that happens at the call is that a *seed* for the new chare is created, which captures the constructor arguments for it; a seed-balancer dynamically moves the seeds among processors for load balancing, until it is scheduled for execution on some processor by executing its constructor, at which point we can assume that the chare has *taken root* there. Chares can obtain their own global IDs (called *proxies* in Charm++), and methods can be invoked asynchronously using these proxies. Parallel programming based on such dynamic creation of individual chares is useful in a variety of situations, including combinatorial search.

For applications in science and engineering, we need a further abstraction: multiple chares may be organized into a collection, and each chare belonging to a collection can be named (and accessed) by an *index*. For example, one may have a one-dimensional array of chares. One can then broadcast method invocations to all the elements of a collection, or to a single named one. These collections are called *chare arrays*. However, they are not limited to be simple arrays. The index structures may define collections that are multi-dimensional sparse structures (e.g., a 6-dimensional array, with only a small subset of possible indices being instantiated as chares). They can also be indexed by other arbitrary indices, such as strings or bit vectors, but such usage is not common in current CSE (Computational Science and Engineering) applications.

A single program may contain multiple chare arrays. These may arise from multiple application modules, or a single module whose algorithm is more naturally expressed in terms of multiple chare arrays. To communicate with chares belonging to a chare array, one must get hold of a "proxy"—an object that stands for (or refers to) the whole collection. A proxy is returned when a chare array is first created. So, the code `A[i].foo(x,y);` specifies asynchronously sending a method invocation for the method `foo` with parameters

---

[3]The model up to this point is similar to the "actor" model developed by Hewitt, Agha, Yonezawa and others, with the possible exception of the idea of an explicit "mailbox" that an *actor* has access to. More important points of departure come in the features described after this point, and in the reliance of Charm++ on its extensive adaptive runtime system.

x,y to the i'th object of a 1-dimensional array referenced via its proxy A. The call immediately returns, while the method invocation is packaged and sent to the processor where the ith element resides.

Chare arrays support reductions and broadcasts over all its elements, but, unlike in MPI-1 or MPI-2, these are both asynchronous non-blocking operations. (MPI-3 standard has now adopted non-blocking collectives.) A broadcast can be initiated by any element or even from other chares not belonging to the target chare array. In our example above, A.foo(z,t) will result in asynchronous invocations of the foo method of all the member chares of A, a broadcast. The system ensures that all the chares belonging to a chare array receive the successive broadcasts in the same sequence. Reductions are carried out via non-blocking "contribute" calls that allow other computations to proceed while the result of the reduction (such as a global sum) is delivered to its intended target, via an entry method invocation or via a general-purpose callback abstraction. In particular, the members of the chare array over which the reduction is being carried out are free to execute other entry methods while the reduction is in progress.

The chares belonging to a chare array are assigned to processors by the runtime system (RTS) as shown in Figure 1.3; the RTS may change this assignment at runtime as needed. A scalable location manager [147] keeps track of which chares are on which processor, resulting in messages being delivered quickly and with low overhead to the right processor.

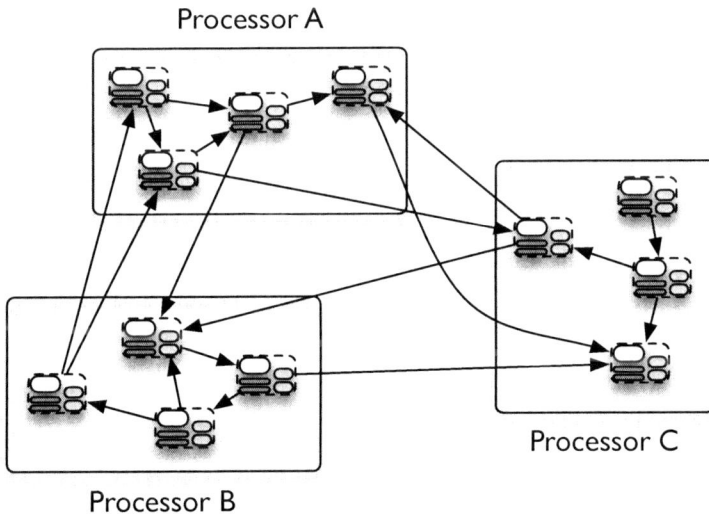

FIGURE 1.3: System view of chares.

1.3   Capabilities of the Adaptive Runtime System

The heart of the Charm++ system, and its signature strength, is its adaptive runtime system. The primary responsibilities of the runtime system are:

1. To decide which objects reside on which processor, when they are created,

2. To schedule (i.e., sequence) the execution of all pending entry method invocations on the targeted chare objects on each processor,

3. To keep track of current location of each chare, in spite of chare migrations, in a scalable and low-overhead manner,

4. To mediate communication between chares by delivering entry method invocations to the correct target object on the processor where it resides. And, finally,

5. To migrate chares across processors, if needed, in response to runtime conditions.

The Charm++ programming model provides much flexibility to the runtime system, in terms of placement of chares on processors, sequencing of method invocations and mediating and intercepting communication between chares. The Charm++ adaptive runtime system (RTS), thus empowered, leverages this flexibility to optimize performance as the program executes. Here, we will briefly discuss its capabilities in balancing load dynamically, tolerating faults, optimizing communications, and managing power.

Dynamic Load Balancing: Charm++ supports a large suite of load balancing strategies. Some of these strategies use measurements of computational loads and communication graph between chares, which the RTS can readily obtain because of its role in scheduling chares and mediating their communication. With Charm++, load balancing can be thought of as a two-phase process: the programmer decomposes the work (and data) into chares. This division does not have to be perfect, i.e., significant variation in the work/size of chares is permissible, since there are typically tens of chares on each processor core. At runtime, the RTS assigns and reassigns chares to individual processors, to attain such goals as better load balancing, and/or minimization of communication volume. As the number of chares is much smaller than the number of underlying data-structures elements (e.g., grid points, or mesh elements), the load balancing decisions are much faster than, say, applying a graph partitioner such as ParMETIS to the entire underlying structure. Occasionally, chares may have to be split or merged to keep their size within a desired range; Charm++ supports dynamic creation and deletion of chare array elements if needed. But this is not needed for most applications, and when it is needed, it is still simpler than a complete repartitioning of the application data structures.

The suite of strategies provided with Charm++ includes some that ignore communication volume and some that consider it. It also includes some strategies that *refine* load balance, by moving a relatively small number of objects from overloaded processors, and other schemes that comprehensively repartition the object graph. For large machines, it includes strategies that optimize placement with respect to the interconnect topology, and hierarchical strategies that significantly reduce decision time. Further, one can write application-specific strategies (or new, general-purpose ones) using a relatively simple plug-in architecture. Also, a *meta-balancer* that examines application characteristics, to choose the appropriate strategy, and decide when to apply it, has been developed recently.

Automatic Checkpointing: Parallel application developers often need to write code for periodically checkpointing the state of their application to the disk. Simulation runs are often long, and need to be broken down into segments that will fit within system-allowed durations; also, hardware failures may cut short an ongoing simulation. Checkpoints allow one to handle such situations without losing much computation. Since Charm++ already has the capability of migrating objects to other processors (with users providing information to optimize the amount of data saved, if needed), the RTS can leverage this capability to "migrate" copies of objects to the file system, along with the state of the runtime system itself. This reduces the burden on the programmer as they do not need to write additional checkpointing code. Further, when restarting, they can use a different number of processors than what was used for the original simulation, e.g., a job that was running on 10,000 cores can be restarted on 9,000 cores! This works automatically for baseline Charm++ programs, and requires little extra programming for programs with user-defined groups and node-groups (Section 1.4).

Fault Tolerance: One can also make a Charm++ application continue to run in spite of individual nodes crashing in the middle of the execution! Charm++ offers multiple alternative schemes for this purpose. The most mature, and probably most useful for applications today, is the double checkpointing scheme, which stores a checkpoint of each object locally and on a buddy node. An automatic failure-detection component checks the "heartbeat" of each node in a scalable and low-overhead manner. When a node fails, the system effects a recovery by automatically and quickly restoring the state of the last checkpoint. How quickly? We have measured restarts in hundreds of milliseconds for a molecular dynamics benchmark on over 64k cores [254]! Even on applications with very large checkpoints, it usually takes no more than a few seconds. One can use spare processors or make do with remaining processors on failure. For large runs, running with a few spares is inexpensive and simplifies the load balancing the system must do after restart.

A more advanced scheme based on *message-logging with parallel restart* has also been developed [35, 34, 169]. With the double-checkpoint scheme (as with any other checkpoint-restart scheme), when a node fails, *all* the nodes must be rolled back to their checkpoints, wasting energy and wasting a lot of computation. With our message-logging schemes, when a node fails, only its objects restart from the checkpoint, while the others wait. The restarting objects can recover in parallel on other processors, thus speeding recovery. It does require storing of messages at the senders, which can add to memory overhead. Many strategies aimed at reducing this overhead have been developed [170]. This scheme is expected to be more important beyond Petascale, when node failures are likely to be frequent.

Power Management: Power, energy and temperature constraints are becoming increasingly important in parallel computing. Charm++, with its introspective runtime system, can help by monitoring core temperatures and power draw, and automatically changing frequencies and voltages. It can rely on its rate-aware load balancers (i.e., strategies that take into account the different speeds of different processors) to optimize either execution time or energy, while satisfying temperature and total-power constraints. As an illustration [214, 192], we were able to reduce cooling energy in a machine room by increasing the A/C thermostat setting; of course, that may lead to some chips overheating. However, the Charm++ runtime system monitored chip temperatures periodically, and lowered the frequencies of chips that were getting too hot, while increasing them if they were cold enough. This creates a load imbalance which would slow the whole application down, as the rest of the processors wait for data from the processor whose frequency was lowered. However, the runtime is able to rebalance load by migrating objects after such frequency changes. These power-related features are available only in experimental versions of Charm++ at the time of this writing, but are expected to be more broadly available in the near future.

Communication Optimizations: The Charm++ runtime system is continually observing the communication patterns of the application, since it is delivering messages to chares. It can replace communication mechanisms based on the observed patterns. For example, algorithms for collective communication can be changed at runtime, between iterations of an application, based on the size of messages, number of nodes and machine parameters [139].

1.4   Extensions to the Basic Model

In Section 1.2 we described the basic Charm++ programming model, consisting of chares, indexed collections (arrays) of chares, asynchronous method invocations, broadcasts and reductions. This baseline description is very useful for developing an intuition about the programming model, and its underlying operational semantics (or execution model). A few important extensions to the base model, which enrich the programming model without changing its essential character, are noted below. These "extensions" are as mature and old as Charm++ itself, and are in common use in applications today.

Supporting "Blocking" Calls: Charm++ supports two additional kinds of methods, specifically tagged as such, that do allow "blocking" calls. They do not block the processor; only the affected entry method is paused, and control is returned to the scheduler. These are called `Structured Dagger` methods and `threaded` methods, as explained below.

A `Structured Dagger` (also abbreviated `sdag`) entry method allows users to define a DAG (directed acyclic graph) between computations within a chare, and asynchronous method invocations expected by the chare. This typically allows one to express the life cycle of a chare object more clearly than a baseline program would. An important statement in the structured-dagger notation is the so-called `when` statement, which specifies (1) that the object is ready to process a particular entry method invocation, and (2) what computation to execute when this method invocation arrives.

Just to give a flavor of how `sdag` code looks like, we present a snippet of code below. This comes from a molecular dynamics example, discussed briefly in the next chapter. But that is not important here; we are just illustrating the structure of `sdag` code. The "run" method of this chare includes a time step loop. In each time step `t`, the `run` method waits for two invocations of coordinates method, and when both are available executes some sequential object methods atomically. The sequential code calculates forces on each set of atoms C1 and C2 due to the other set of atoms, and sends the resultant forces back. Since this is not usual C++ code, sdag entries are specified in a separate file, which is translated into C++ code. One can thus think of `sdag` code as a *script* for describing data-dependent behavior of a chare. Typically, it describes the entire life cycle of a chare, as signified by the name, "run" method, in this particular case.

```
1  entry void run() {
2    for (t=0; t<steps; t++) {
3      when coordinates(vector <Atom> C1),
4           coordinates(vector <Atom> C2)
5        serial {
6          calculateInteractions(C1, C2);
7          sendForcesBack();
8        }
9    }
10 };
```

When a *threaded method* is invoked the runtime system creates a lightweight user-level thread and starts a method invocation inside this thread. A threaded entry method can block waiting for a *future* [95], or for a return value from a *synchronous* method invocation. Correspondingly, the system allows users to define entry methods that return a value, as well as a simple future abstraction. One can create a `future`, set value to it or access value from it (which is a blocking call). If a thread tries to access the value of a *future*, and the value is not set yet, the thread is blocked, and control is transferred to the Charm++ scheduler. Later, when the value is set, the thread is added back to the scheduler's queue, so it can be resumed in its turn.

Array Sections: A subset of chares belonging to a chare array can be organized into a *section*, somewhat like an MPI sub-communicator. One can invoke broadcasts and reductions over sections as well. The system organizes efficient spanning trees over the subset of processors that house elements belonging to a section. It ensures that the broadcasts and reductions are carried out correctly even when element chares migrate to other processors, and reorganizes the spanning trees periodically, typically after a load balancing phase.

Processor-awareness: Another extension has to do with awareness of processors by the programmer. In the model described so far, there is no need for the programmer to know anything about the processors, including which processor is the current location of a particular object. However, there are some situations in which an "escape valve" into processor-aware programming is needed. This is especially true for libraries, or performance oriented optimizations. For example, many objects on the same processor may request the same remote data; it makes sense in this situation to use a shared processor-level software cache. Requests for remote data can go via this cache object, and if requested data was already obtained due to another chare's request, unnecessary remote communication is avoided. For such purposes, Charm++ provides a construct called chare-group. Just like an array of chare objects, a chare-group is a collection of chares. However, (1) there is exactly one member mapped to each processor, and (2) unlike regular chares, chare group members are allowed to provide public methods that are invoked directly, without needing the packaging and scheduling of method invocations. Also, given the group ID, the system provides a function that returns a regular C++ pointer to the local (branch) chare of the group. With these two features, chares can communicate using low-overhead function calls with the member ("branch") of a group on their processor. Note that such group objects also allow additional data sharing mechanisms [222] beyond the read-only variables mentioned earlier.

So far, we intentionally left the notion of what we mean by a "processor" only loosely defined. In Charm++, for processor-aware programming, there is a notion of PE (processing element). A PE corresponds to a single scheduler instance; a Charm++ application may associate a PE with a hardware thread, a core or a whole or a part of a multicore node, based on command-line options. If a PE includes multiple hardware resources (say cores), the parallelism within a PE is managed by the user orthogonally, by using pthreads, openMP or Charm++'s own task library (called `CkLoop`). Associating a PE with a hardware thread is a common practice in current Charm++ applications, and it obviates the need to deal with an additional level of parallelism, so we will assume this in our description.

Of course, the group construct and other such low-level features should be used sparingly. As a design guideline, one should strive to avoid using processor-aware programming as much as possible, and push it into low-level libraries when needed. The example in the above paragraph, involving multiple requests for remote data, is a common enough feature that a new library, CkCache, has been developed as a common library for use by multiple applications. The system libraries for implementing asynchronous reductions are another example. Although one could implement a spanning tree over all the chares of a chare-array, it is much more efficient to do a processor (and node) based spanning tree, collecting inputs from all the local chares first.

Since objects may be migrated by the runtime system to other processors, Charm++ also supports a special callback method that gets called after the object has been reincarnated on another processor; this can be used to update any processor-specific information, such as pointers to local branches of groups, stored by the objects.

1.5   Charm++ Ecosystem

Charm++ is a mature and stable parallel programming system. Thanks to the popularity of applications such as NAMD, it is used by tens of thousands of users worldwide. (The biomolecular simulation code NAMD, described in Chapter 4, has 45,000 registered users, as of December 2012.) Charm++ is available on most national supercomputer installations in the US. Charm++ runs on almost all the parallel computer types that are widely known, including Cray machines, IBM Blue Gene series machines, Linux Clusters, Windows clusters, etc. It supports multiple network types including proprietary networks on supercomputers, as well as commodity networks including Ethernet and Infiniband. Charm++ is regression-tested via a nightly build system on dozens of combinations of compilers, operating systems, processor families and interconnection networks.

The maturity of Charm++ is also reflected in the ecosystem of program development and analysis tools available for it. *Projections* is a sophisticated performance visualization and analysis tool. *CharmDebug* is a more recent and highly sophisticated debugging tool. In addition, the *LiveViz* library can be used to collect application or performance data during application run and visualize it as the program is running. The CCS (Converse Client-Server) library that underlies LiveViz also allows one to develop interactive parallel applications, whereby queries or messages can be injected into a running computation, either to examine specific attributes of a running simulation, or to effect changes in the execution of the application.

There are several online resources for learning Charm++ and working with it. The software, manuals, tutorials and presentations are available at http://charm.cs.illinois.edu. An active mailing list (charm@cs.illinois.edu) is used for reporting bugs and discussing programming issues and upcoming features. There is an annual workshop on Charm++ and its applications in Urbana, Illinois; the presentations from the workshop (starting in the year 2002), most including the video of the presentations, are also available online.

1.6   Other Languages in the Charm++ Family

Charm++ is just one instance of a broader programming paradigm based on message-driven execution, migratable work and data-units, and an introspective and adaptive runtime system. Although Charm++ is the earliest member of this family of programming languages there are a few others that we have developed that deserve a mention here. All of these are built on top of Charm++, as Charm++ turns out to be an excellent backend for developing new abstractions within this broad paradigm.

XMAPP is the name we have chosen for the abstract programming model that underlies Charm++ as well as all the other languages described below. XMAPP is characterized by a few defining attributes:

- Over-decomposition: the interacting entities, be they units of work or units of data (or a mix of the two, as in Charm++), into which the computation is decomposed by the programmer in such models are independent of the number of processors, and typically their number is much larger than the number of processors.

- Processor-independence: the interaction/communication between entities is in terms of names of those entities and not in terms of processors.

- Migratability: these entities can be migrated across processors during program execution, either by the runtime system, or the application itself, or both.

- Asynchrony: collectives and other communication-related operations are designed so that their implementations do not block the processor.

- Adaptivity: the runtime system takes responsibility of balancing load by leveraging its ability to migrate objects.

Adaptive MPI (AMPI) is an implementation of the MPI standard on top of Charm++. In MPI, the computation is expressed as a collection of processes that send and receive messages among themselves. With AMPI, each MPI process is implemented as a user level thread. These threads are embedded inside

Charm++ objects, and are designed to be migratable across processors with their own stack. As with Charm++, multiple "processes" (i.e., MPI ranks) are typically mapped to a single core. Standard MPI calls, such as those for receiving messages, provide natural points to allow context switching among threads within a core, thus avoiding complexities of preemptive context switching. AMPI programs have shown to have comparable performance (somewhat slower for fine-grained messaging, but comparable for most applications) as the corresponding MPI program, even when no AMPI-specific features are being used. Those features, such as over-decomposition (and adaptive overlap of communication with computation), asynchronous collectives, load balancing and fault tolerance, provide the motivation for using AMPI instead of plain MPI implementations.

MSA (Multiphase Shared Arrays) [48, 173] is a mini-language on top of Charm++ that supports the notion of disciplined shared memory programming. It is a partitioned global address space (PGAS) language. The program here consists of a set of migratable threads and a set of data arrays. The data arrays are partitioned into user-defined "pages," which again are migratable data units implemented as chares. The main point of departure for the language is the notion of *access modes*. Each array may be in one of the few possible modes, such as "read-only" or "accumulate." All the threads must collectively synchronize to switch the mode of an array. This model is shown to avoid all data races, and yet captures a very large fraction of use cases where one would find shared global data useful.

Charisma [104] is a language that allows elegant expression of applications or modules that exhibit a static data-flow pattern. The computation is divided into chares. If the chares exchange the same set of messages (with different lengths and contents, to be sure) in every iteration, one can express the lifecycle of entire collections of chares in a simple script-like notation, where entry-methods are seen to publish and subscribe to tagged data.

Charj [173] is a compiler supported language that provides the same abstractions as Charm++ combined with MSA. With compiler supported static analysis, Charj provides a more convenient and elegant syntax, automatic generation of serialization code and several other optimizations based on static analysis. Charj is an experimental or research language at the current time.

1.7 Historical Notes

The precursors to Charm++ (the "Chare Kernel") developed by us were aimed at combinatorial search applications, and at supporting parallel functional and logical programming. However, once we turned our attention to science and engineering applications, we decided to mold our abstractions based on the needs of full-fledged and diverse applications. The first two applications examined were fluid dynamics [92] and biomolecular simulations [112]. These two applications, along with many small examples, and parallel algorithms such as the Fast multipole algorithm, histogram-based sorting [116, 137] adequately demonstrated to us that our approach was avoiding the trap of being too specialized. This was especially true because we considered full-fledged applications, in addition to kernels or isolated algorithms. We thought that only by immersing ourselves in the nitty-gritty of developing a full-fledged application would we be able to weigh the importance and relevance of alternative abstractions, and capabilities.

This position and approach towards development of abstractions were explicitly written down in a position paper around 1994 [114]. The biomolecular simulation program NAMD funded by NIH (and NSF, in the early days, under the "Grand Challenge Application Groups" program) provided us a good opportunity to practice and test this approach. NAMD was developed in collaboration with Klaus Schulten, a biophysicist with a computational orientation, and Bob Skeel, a numerical analyst, both professors at the University of Illinois.

1.8   Conclusion

We believe that Charm++ and the underlying XMAPP abstract programming model constitute an approach that is ready to deal with the upcoming challenges in parallel computing, arising from increasingly complex hardware and increasingly sophisticated applications. It appears to us that the basic concepts in XMAPP are going to have to be inexorably adopted by the community, whichever language they choose to use in the future. So, why not Charm++? Charm++ itself is a production-quality system that has demonstrated its capabilities in improving programmer productivity and in attaining high scalability on a wide variety of the parallel applications in science and engineering, as demonstrated by this book. Some applications have demonstrated scaling beyond half a million processor cores by now.

To simplify and ease adoption, Charm++ supports interoperability with MPI: some modules can be written in regular MPI, while others can be based on Charm++ or AMPI (or any of the other mini-languages in the Charm++ family). We invite the readers to experiment with this approach by writing modules of their applications in it, or by using an existing Charm++ library in their MPI application, or testing it by developing an isolated algorithm using it, and then possibly moving on to developing entire applications using Charm++, and reap the productivity and performance benefits.

Chapter 2

Designing Charm++ Programs

Laxmikant V. Kale

Department of Computer Science, University of Illinois at Urbana-Champaign

We learned about the design philosophy, and basic concepts in Charm++, as well as its features and benefits in Chapter 1. In this chapter, we will review the process of designing Charm++-based applications, and discuss the design issues involved. Specifically, we will illustrate how to use individual features of Charm++, through a series of examples.

2.1 Simple Stencil: Using Over-Decomposition and Selecting Grainsize

Let us first consider the process of developing a parallel implementation of a simple stencil code. Through this example, we will illustrate how to specify over-decomposition in practice, and how to make practical grainsize decisions. We will use the same example to show how to extract automatic communication-computation overlap (which is a benefit of the Charm++

model), and show how easy it is to exploit features such as load balancing and fault tolerance.

Imagine that the data that we wish to deal with is represented by a three-dimensional grid. Further, the computation we wish to carry out for each cell involves using values from the neighboring cells in the grid. To make it concrete, let us consider a 7-point stencil: calculating the new value of a cell located at location [x,y,z] in the grid requires values of the cells whose coordinates differ by just one in exactly one of the dimensions involved. Other than this somewhat abstract description of the pattern of data communication needs, we will omit the rest of the details of the computation to keep the description simple. Suffice it to say that this communication pattern arises in many science/engineering simulations, such as those involving solving the Poisson equation, or fluid dynamics computations.

How should the three-dimensional grid be decomposed among the processors? For simplicity of analysis, let us assume that the grid has a cubical aspect ratio. That is, the number of cells along each of the dimensions is the same. If one is designing the application with a traditional programming model, such as MPI, one can consider several options: for example, the data grid may be divided into horizontal slabs, with one slab assigned to each processor. However, with a sufficiently large number of processors, it is known that the communication volume (i.e., the amount of data being communicated in total) is smaller if one divides the grid into cubes, and assigns one cube to each processor.

Again, for the purpose of simplicity, let us assume that there is only one processor core on each node. So, when we say "processor" in the above paragraph, we simply mean a node. We will return to the issue of multicore nodes in the context of Charm++ a little bit later.

Notice that the above strategy in case of the simple MPI programs requires us to request a cubic number of processors. In addition, often the stencil codes keep the number of cells on each processor exactly equal by requiring that the total grid size along each dimension be a power of 2. This also helps keep the code simple, in particular in the part where each processor needs to calculate which portion of the global grid it owns. However, this decision now has an unintended consequence: the number of processors has to be a cube of a power of 2! Alternatively, one can write clever MPI code that can utilize a MxNxK processor grid with some careful decomposition along each dimension, or even write a much more sophisticated (and complex) multi-block code in MPI.

2.1.1   Grainsize Decisions

When we think about the same problem in Charm++, we should still choose a cubic decomposition. The grid is represented by a three-dimensional array of chare objects. Each object communicates with its neighboring objects so as to send the boundary data that they need to complete the stencil computation. However, we pretty much ignore the number of processors in deciding the decomposition. Below, we will describe several considerations one may use in deciding how large each cube should be. However, the important thing to note is that the *number of processors* is not a primary consideration.

A basic guideline in deciding how large each cube (and in general, each chare object) should be is: *make it as small as possible, but large enough that its computation is significantly larger than the overhead of scheduling all the method invocations it must handle.* On current machines, typically, the overhead of scheduling a single method invocation is hundreds of nanoseconds. There may be memory allocation overheads that make it slightly larger. CPU overhead for remote communication is of the order of a microsecond at most in modern machines. In any case, since each one of the cubes must send and receive about 6 asynchronous method invocations each cycle, if we make the computation larger than, say, 100 microseconds, that should be adequate. As we will see below, exact determination of the optimal grainsize is not needed.

Other considerations such as memory usage may push the application developers to use a larger grainsize than that. In any case, a common approach among Charm++ application developers is to parameterize the choice of grainsize, and to determine the size to use in production runs experimentally.

In applications that use a weak-scaling approach, where the amount of data used on each processor remains roughly constant, these processor-independent guidelines can be translated into processor dependent terms. For example, an application developer may say that this application behaves reasonably well when there are between 20 and 40 objects per processor. It should be understood that the basic consideration is still the size of each object, and the number of objects per processor (often called "the virtualization ratio") is a secondary, derived metric.

So, the important point to remember: in most applications, it is not necessary to precisely decide the grainsize. As long as it is sufficiently large to amortize the overheads, and small enough to make several chares on each processor available to the adaptive runtime system, there is a range of grainsize values that yield similar, and close to optimal, performance.

If dynamic load balancing is necessary for the application, it helps to have more than 10 objects per processor. It also is important in that situation that no single object be too large. If a single object is larger than the average load per processor, for example, there is nothing a load balancer can do to reduce the execution time below a single object's time. It is usually easy to meet the constraints on the average grainsize (described in the paragraphs above), as well as this maximum grainsize constraint, because of the broad range of grainsizes that satisfy those constraints. In some rare applications, it may become necessary to split objects that become too large computationally as the application evolves. Charm++ supports dynamic insertion of new objects for this purpose.

Returning to our simple stencil computation, we have now defined the three-dimensional array of objects that is being distributed to processors under

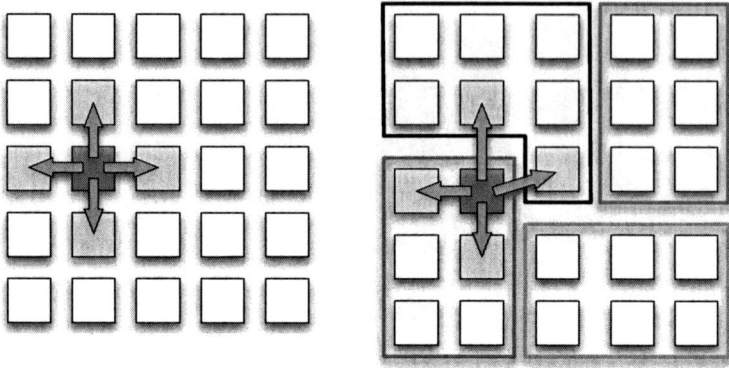

FIGURE 2.1: Chares for a 2D Stencil computation and their mapping to physical processors.

the control of the runtime system. If we choose, we can describe a mapping (a user-defined assignment of objects to processors). We may specify the mapping to be immutable or just an initial mapping that the runtime system can change as the application evolves. Programmers control which type of mapping and dynamic reassignment they desire.

Even for this simple, regular program we can see several benefits in a Charm++-based design. The program will run fine on any given number of processors. It will just distribute all objects on available processors (See Figure 2.1, with a 2-D example, for simplicity). Sure, some processors may get one more object than other processors, but this "quantization" error is certainly acceptable as it is typically less than 5-10 % penalty, and applies *only when* the number of processors does not evenly divide the number of objects. Typically, cache performance of the code improves because of the blocking effect of smaller objects, leading to improved performance compared with having just one chunk of each processor. Communication is naturally overlapped with computations; while some chunks are waiting for their communication, other chunks can compute, the interleaving occurring without programmer intervention, because of the message-driven execution model.

How does our execution model work for this application? Looking a bit under the hood, on each processor we have a separate scheduler (See Figure 1.2 and the associated description in Chapter 1), working with a pool of method invocations waiting to be executed. It picks one of them, delivers it to the targeted object, and when (only when) the object returns control back to the scheduler, it repeats the cycle. Notice how this naturally interleaves execution of objects, adaptively and automatically overlapping communication and computation.

2.1.2   Multicore Nodes

Let us return to the issue of **multicore nodes** now. The programmer has multiple options for dealing with this. One is to assert that there is one "PE" associated with each core (or even, with each hardware thread). One still expects there to be multiple objects on each PE, and the model works as described above, except (1) communication within a physical node is fast because the RTS uses shared memory for implementing it. (2) When needed, the application itself can use the shared memory in specific ways to optimize performance. As a simple example, Charm++ supports read-only variables, for whom only one copy needs to be made on each node.

A second option is to treat a collection of cores (say a NUMA domain, i.e., a set of core sharing some level of cache and/or a memory controller) as a single PE. In this case, the decomposition may employ chunkier objects. Now, each PE may utilize multiple cores associated with it by using many possible constructs: for example, OpenMP, or Charm++'s internal CkLoop construct that can use the cores via their associated hardware threads.

With the second mode, message sizes are larger, and the memory locked up in ghost regions (for our stencil example) is proportionately smaller. The former achieves better overlap of communication and computation, and *may* have cache performance benefits depending on patterns of data accesses. Which method is better depends on many application dependent factors, and the programmer can make the right choice. The essential characteristics of the program are still unchanged with either option.

2.1.3   Migrating Chares, Load Balancing and Fault Tolerance

For various reasons, including load balancing, we may want the chares to be able to migrate from one processor to another during the execution. To facilitate this, the programmer must provide the system a bit of additional information: how to serialize (i.e., pack and unpack) the data of the object. This is done in Charm++ using a powerful and flexible PUP (pack-and-unpack) framework. For every chare class that you wish to make migratable, you must declare a PUP method. This method simply enumerates the variables of the object, with a few notations to indicate sizes where they cannot be inferred. The system will call the same PUP method of a chare for finding the size of the chare, for serializing its contents, and for reconstructing (deserializing) it from a packed message on the other processor.

Once the chares are made migratable, many new benefits follow. For example, one can migrate "work to data" at will. Better still, you can let the system do some load balancing for you. In this simple application, all chares have equal amount of work (except for quantization effect because the number of rows and columns may not be divisible by the corresponding dimensions of the chare array). Yet, suppose, this application is running on a cluster consisting of processors of differing speeds.[1] You can now just turn on load balancing, insert a call to balance load only after first few iterations (so the system has the load data to base its decisions on) and presto: the chares will be allocated in proportion to the speed of the processors, with faster processors getting proportionately more work. This is achieved using chare migratability and the RTS's ability to keep track of processor speeds and chare loads (as well as chare communication patterns).

In fact, suppose we add some kind of dynamic load imbalance to the problem, say, by adding particles to our simulation. The only additional change you have to do is to make the call to balance load every few iterations, instead of doing it at the very beginning.

Now, suppose, you want to checkpoint the state of the program periodically, so that you can start from the last checkpoint if the program runs out of allocated time before it finishes. Well, since the system knows how to migrate your chares to other processors, it can migrate them to disk! The RTS has to perform some complicated maneuvering to ensure its own state is stored accurately when a checkpoint is taken; but for the application programmer, the code involved is simply calling `CkStartCheckpoint(char * dir, const CkCallback& cb)` collectively from every chare, every certain number of iterations. Additionally, for most applications, you can restart using the saved files on a different number of processors. The set of original chare objects is just scattered on a different number of processors.

This kind of disk-based checkpoint is not quite true "fault-tolerance" because it does not include automatic restart. But Charm++ also supports true fault tolerance strategy: it is called *in-memory double checkpointing.* In this case, you would have to provide simple link-time and compile-time options (the details of which can be found in the Charm++ manual and tutorials). After that, in your code, you simply add a different call to checkpoint (instead of the disk-based checkpoint call mentioned above). To test this, suppose this program is running on a cluster of workstations. At any random point in time, try just killing the application process on one of the processors (in Linux, just `kill -9 PID`). The system will detect that one of the processes has died, retrieve the in-memory checkpoint, roll back to it and continue execution from it in less than a second! Of course, on proprietary machines such as Cray and IBM installations, or clusters running under a job scheduler, there is a problem: the job schedulers running on today's supercomputers will simply kill a job if one node dies. However, as the schedulers get more sophisticated, they will support such fault tolerance strategies; this is starting to happen on some of the current supercomputers.

---

[1]Years ago, we had this situation at the CSAR (Center for Simulation of Advanced Rockets): we had a cluster, and then new faster nodes were bought and added to the cluster. Of course, most MPI jobs confined themselves to either the new partition or the old partition entirely. However, Charm++ jobs were able to run on the combined system, adjusting to the speed heterogeneity with its load balancers.[29]

2.2 Multi-Physics Modules Using Multiple Chare Arrays

One of the powerful techniques one can use in designing parallel Charm++ programs is to use multiple chare arrays. This can be used to eliminate unnecessary "coupling"[2] between software modules, and to promote modularity as well as facilitate collaborative development of software.

As a first simple example, consider a hypothetical simulation that involves:

- a solid modeling component (e.g., structural dynamics) that requires the use of unstructured (say, tetrahedral) meshes, which are partitioned using a program such as ParMETIS or Scotch.

- a fluid modeling component that uses structured grids, partitioned using a different partitioner.

In a traditional MPI application, one either uses spatial decomposition, dividing the set of processors between the two components, and thus sacrificing efficiencies of using idle times of one module for advancing the computation of the other or, alternatively, one can decompose both computations on the same set of processors (Figure 2.2). However, with the latter approach, multiple artificial couplings develop. The number of partitions of solids must be the same as that for fluids (equal to the number of processors). Further, the partitions assigned to, say, the 200th processor for the fluids' solids are simply brought together because their partitioners called them 200th! Of course, more sophisticated means can be used, but at a significant cost to the programmer. With Charm++, the solids can be partitioned into n partitions, while the fluid data is partitioned in m components (i.e., m and n are not necessarily equal). Further, the runtime controls the mapping and may bring together the pieces that communicate with each other the most.

Another example of the use of multiple chare arrays is provided by NAMD (See Chapter 4), which uses separate sets of chares for atoms, for pairwise explicit force calculations, for bonded force calculations and several chare arrays for the particle-mesh Ewald sub-algorithm. Even more dramatic is the use of over a dozen Chare arrays in OPENATOM (Chapter 5) for representation of the electronic structure, as well as for intermediate parallel data structures.

Multiple chare arrays are also useful in building highly reusable libraries. For example, a sorting library can use a chare array, that can be "bound" to the application's chare array (so that the corresponding elements migrate together and are kept on the same common processor as each other), and can carry out sorting of the data supplied by the application chares.

---

[2]In software engineering terminology, such coupling is considered detrimental to good software structure.

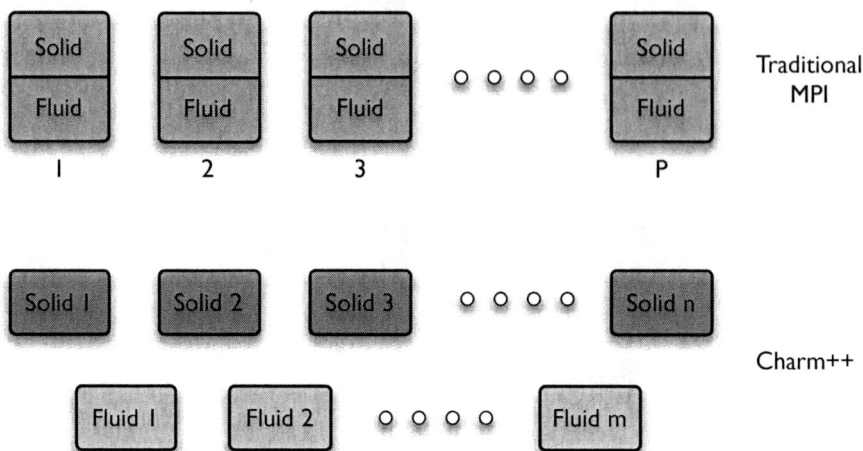

FIGURE 2.2: Decomposition in traditional MPI versus Charm++.

2.2.1   LeanMD

To illustrate the process of using multiple chare arrays for improving modularity and increasing parallelism, let us consider *LeanMD*. LeanMD is a mini-app meant to mimic the structure of the dominant aspects of NAMD, and illustrates this usage of multiple arrays in a much simpler context with a few hundred lines of code [120, 119, 166].

The simulation consists of a set of particles (such as a set of inert gas atoms). For simplicity, we will assume that the particles are confined to a 2-dimensional periodic box. Each particle experiences a force due to every other particle (for example, Van der Waal's force); however, the force decays sharply over distance (we assume), and so we can ignore forces due to all atoms beyond a certain cut-off distance. In each time step, we calculate and add up the forces experienced by each atom, and use them to calculate new accelerations, velocities and positions of each particle using standard Newtonian physics.

The first design decision is how to decompose the data into objects (chares). A simple idea is linear decomposition: the first k particles go the 0'th chare, and so on. But then, every pair of chares will have to exchange particles to see any of them exert forces on the other. Instead, we decompose particles based on their coordinates into boxes. If we choose the size of each box to be slightly larger than the cutoff distance, each box will need particles from only the neighboring 8 boxes. With MPI, at this point in the design, you will start thinking about a multi-block code, and writing the coordination code for managing multiple boxes on a processor. But with Charm++, you just think of each box as a virtual processor (i.e., a chare) by itself, and describe its life cycle in a coherent code, without thinking about the physical processor. Also, at this point, you will do a simple grainsize analysis to decide if it is worthwhile using a finer decomposition (say, sizes of boxes being about half of the cutoff distance, and allowing interactions with boxes in a larger neighborhood). For the simplicity of this example let us stay with the basic decomposition. This immediately requires us to program for exchange of particles after every time step so that each particle is in the correct box. (We can reduce that communication by increasing the box size slightly and doing particle exchange after multiple steps.)

The interesting issue is the scheme for calculating forces. Since neighboring boxes must interact, a straightforward idea is to send the particles of a box to all the neighboring boxes, get the particles from the neighboring boxes and calculate and add up forces on your own particles from all the relevant particles (that you now have). However, because of Newton's third law, we need to calculate forces between each pair of atoms (and by extension, each pair of boxes) only once; so we are duplicating the force computation work. Since this work is known to be over 90% of the overall computation, we wish to avoid the duplication.

One could design complicated schemes for deciding which box calculates forces for which neighbor. But a simpler alternative is presented by our ability to use multiple chare arrays. For each pair of neighboring boxes, we postulate an *interaction* object (see Figure 2.3). These interaction objects can be organized into a chare array. A good index structure for this array is to use a 4-dimensional sparse array. The interactions between box (4,8) and (4,7) is calculated by an interaction object with index (4,7,4,8). We sequence the two indices in lexicographic order, so as to avoid duplication (so that we do not think of this interaction object as (4,8,4,7) as well). As another benefit,

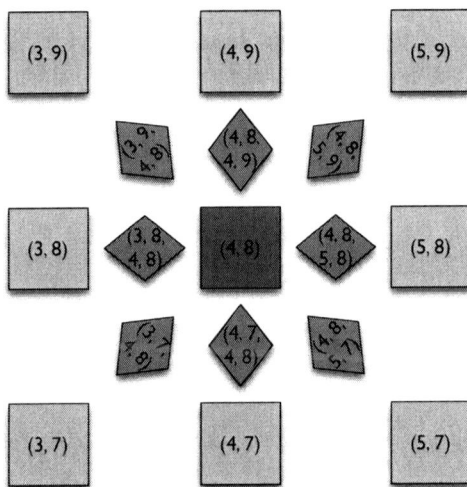

FIGURE 2.3: Use of interaction objects for force calculations in LeanMD.

the interactions between two boxes can now be calculated on a third processor, different than the processors where the boxes live, if the load balancer so desires. This idea (of force computations on a different processor), which was later *proposed* independently by multiple researchers [64, 223, 26], was used naturally in the original NAMD design [121] because we were thinking in terms of object-based decomposition.

Now that the design is done, and especially with factoring of the code between the two types of objects, the code becomes relatively simple. The life cycle of each object is very cleanly expressed: a box object repeatedly broadcasts its coordinates to its associated interaction objects, receives forces from them, adds them up and updates positions of each of its particles. It then (periodically) sends particles that have moved out of its bounds to the neighboring boxes. No particles move so fast as to cross more than one box boundary; or else you are using too large a time step. The almost-real pseudocode in Figure 2.4 illustrates the box's code. We omit the detailed syntax, as well as details of initialization, to avoid getting into a tutorial discussion of syntax, which is out of scope for this book. The complete example code, with a 3-dimensional simulation, is available publicly [120].

The interaction object's life cycle is even simpler: in each iteration it waits for particles from the boxes that are connected with it, calculates forces and sends them back to the two boxes. We showed that code in Chapter 1. We just need to add the two lines for load balancing and fault-tolerance (based

```
1  array [2D] Box {
2  ...
3    entry void run() { // the sdag entry method of a box
4      for (t=0; t<steps; t++) {
5        myInteractions.coordinates(C);
6        // broadcast coordinates to the section comprising
7        // interaction objects that need my coordinates
8
9        // forces received via a section reduction
10       when forces(vector <Force> f)
11         serial { integrate(f); } // and update coordinates C
12       send particles to neighbors;
13       for (i=0; i<numNeighbors; i++) {
14         when moveAtoms(vector <Atoms> A)
15           serial { mergein(A,C); }
16       }
17       if (t%M == 0) { BalanceLoad(); when doneBalancing() {} }
18       if (t%N == 0) { CkStartMemCheckpoint(..); when ckptDone
                         () {} }
19     }
20   };
21  ...
22  }
```

FIGURE 2.4: Skeletal code for the box class.

```
1  array [4D] Interaction {
2  // Each Interaction object is a member of a 4D chare array
3  ...
4    entry void run() {
5      for (t=0; t<steps; t++) {
6        when coordinates(vector <Atom> C1),
7             coordinates(vector <Atom> C2)
8          serial { calculateInteractions(C1, C2);
9                   sendForcesBack();}
10       if (t%M == 0) { BalanceLoad(); when doneBalancing() {} }
11       if (t%N == 0) { CkStartMemCheckpoint(..); when ckDone()
            {} }
12     }
13   };
14 }
```

FIGURE 2.5: Skeletal code for the interaction class.

on in-memory checkpoint with automatic restart) to that code to match the box's code above.

Of course, the sequential details of interaction and integration, and some parallel initializations, must be coded by the user. The rest of the coordination of which boxes and interface objects live on which processor, how to sequence their execution to increase the overlap and so on are left to the runtime system. By writing simple PUP routines as described above the code can do load balancing, automatic fault tolerance (and/or checkpointing to disk), with just a couple of additional lines of code, as shown in Figure 2.5.

2.3 SAMR: Chare Arrays with Dynamic Insertion and Flexible Indices

The stencil example above showed that the chares can be organized into two-dimensional or three-dimensional arrays. The LeanMD example uses a two-dimensional and another four-dimensional array of chares. However, the chares can be organized into even more sophisticated and general indexing structures. In particular, one can use bit vectors as indices. In addition, one can insert and delete elements (i.e., individual chare objects) from arrays of chares, and this can be done dynamically as the computation evolves.

We will illustrate both these features using structured adaptive mesh refinement (SAMR) as an example. SAMR is used in physical simulations where the degree of spatial resolution needed varies significantly from region to region. In one particular formulation that will be our focus, the region is organized as an octree. For the sake of simplicity, let us consider a 2-dimensional example, where we will use a quad-tree instead. The leaves of the tree represent regions that are being explicitly simulated, using a structured grid (i.e., a mesh). The internal nodes in the tree represent regions that have been adaptively refined.

In a simple scenario, computation may begin with a tree of uniform depth. Assume that each leaf has a 512x512 chunk of data. As the simulation evolves, the numerics might indicate that some regions represented by some of the leaves need higher resolution, and so need to be refined. With refinement, the 512x512 grid of the leaf needs to become a 1024x1024 grid. This is accomplished by creating 4 more leaves which become children of the leaf being refined. Although each region can only be refined once during a single time step, over a series of time steps some regions can get very deeply refined, so the maximum and minimum depth of the tree can be widely different. Also, as simulation progresses, some regions may not need the resolution we currently have. If all the 4 children (which happened to be leaves) of an internal node N wish to be coarsened, they can be absorbed in N, thus deleting the 4 leaves.

Charm++ provides a natural way of representing this computation. Each node of the quad-tree, including the leaves, can be implemented as a chare belonging to a single chare array. Each element of this chare array is represented by a bit vector index. In the current version of Charm++, array indices can be 128 bits long. We create a natural indexing scheme that assigns a unique index to each node of the tree as follows: We reserve 8 bits to encode the depth of the node (the depth of the root being 0), and the remaining bits to encode the branches one takes from the root to get to the node. Thus a node with index (d, b) has 4 children, (d+1, b00), (d+1, b01), (d+1, b10) and (d+1, b11), encoded in the 128 bits in the obvious way, with the unused bits at the tail end being set to 0. Figure 2.6 illustrates this indexing scheme, using a smaller number of bits for ease of illustration. Note that the rightmost bits that are not pertinent to the scheme (and are set to 0 for a canonical representation) are omitted by showing them as blanks in the figure.

With this indexing, a chare can find the index of its parent and the neighbors (for boundary-exchanges) by simple local operations based on the knowl-

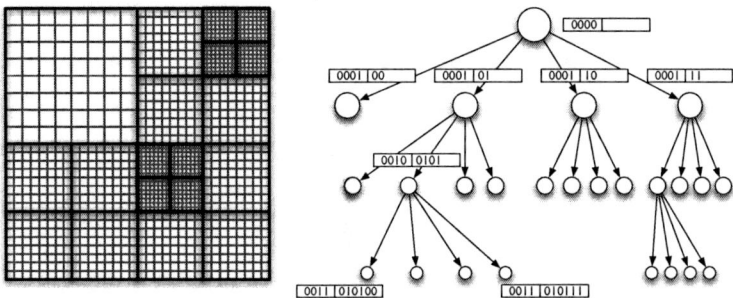

FIGURE 2.6: BitVector indexing of chares for AMR.

edge of its own index. Further, if a chare decides to refine, it can locally calculate indices of the 4 new chares it must create and insert into the chare array, as described in the above paragraphs. The Charm++ load balancer decides on which processor each chare resides based on predicted loads, while a built-in scalable location manager handles delivery of messages directed at specific indices to the correct processor.

An additional benefit of the Charm++ model is that the code is written from the point of view of each leaf, i.e., a block of uniformly refined mesh, along with much simpler code for the chares corresponding to the internal nodes of the tree. This is unlike typical codes for the same purpose in MPI, which must be written from the point of view of a *processor*, where multiple blocks are housed.

A specific implementation of the 2-D SAMR is described in a recent paper [142]. This paper also illustrates how SAMR codes can benefit from the support for asynchrony in Charm++ by eliminating multiple synchronizations in typical re-meshing steps in such codes.

2.4 Combinatorial Search: Task Parallelism

The phrase "task parallelism" is used by different researchers in different senses. Some researchers use it to simply connote any use of message-driven execution. A when block (see Chapter 1) in Structured Dagger is ready to execute as soon as the object has reached the statement, and the relevant messages have arrived. This *task* is then kept in the *local* scheduler's queue of ready tasks from where tasks are picked up one at a time, non-preemptively. However, this is not what we mean by task parallelism here. The pattern we are describing has been called "agenda parallelism," and it occurs in combinatorial search and divide-and-conquer applications.

As an example, consider the problem of finding a k-coloring for a given graph. One can begin with a state in which no vertex of the graph is colored; when you consider each state, you select one of the vertices to color, color it with each of the possible colors that are consistent with the constraint (that no two neighboring vertices should have the same color) and thus create several child states. This then defines a search tree.

How can one parallelize this application using Charm++? This is facilitated by Charm++'s support for dynamic creation of chares. The graph itself does not change during the execution, and so can be represented as a read-only variable/structure. Each chare's state consists of a table of vertices that have already been colored and the assigned colors for those vertices. When it starts execution, each chare heuristically selects an uncolored vertex and fires off a new chare corresponding to each possible color that can be assigned to the selected vertex, sending the color-map as a constructor message. Sophisticated heuristics are used to ensure that the color assignment does not create an easy-to-infer infeasibility, and for selecting a good vertex to color next.

One must exercise reasonable *grainsize* control in this application to avoid excessive overhead on one hand, and serialization on the other hand. Again, the guiding principle is that no single chare should be too large (i.e., its computational work should be substantially smaller than the average work per chare), and the average work for each chare should be significantly larger than the overhead of scheduling and load balancing it. A simple strategy is to decide to explore a state sequentially when the number of vertices that remain to be colored falls below a threshold. More sophisticated strategies can be used, as described in [118].

How do we stop such a computation? It is not adequate to specify that once a chare finds a solution, it will print the solution and call `CkExit()`. For one thing, you may be interested in all the solutions; in any case, you must also cater to the case when there are no solutions. Both of these challenges can be resolved by using Charm++'s built-in quiescence detection library. It employs an algorithm that runs in the background, and reports via a callback when no computations are executing, and no messages are in transit.

Interestingly, this kind of dynamic creation of tasks can co-exist (in a single application) with more structured and iterative computations typically expressed with chare arrays. The message-driven execution combined with load balancing capabilities ensure that such an application is feasible and will work efficiently. At the most, depending on the context, it may require a more specialized set of load balancers, since Charm++ uses separate balancers for such dynamically created tasks.

2.5   Other Features and Design Considerations

Let us consider some additional features and discuss how and under what conditions are they desirable.

Priorities: Recall that there are potentially multiple "messages" (i.e., asynchronous method invocations and ready threads) awaiting execution in the scheduler's pool on any PE (processor). By default, the system executes them in FIFO (first-in-first-out) order. However, in some situations the programmer may wish to influence this order. This can be accomplished by associating a priority with the method invocations (again, consult the manuals for the details of how to do this). One can also declare entry methods, and therefore all invocations of them, to be "expedited," in which case they bypass the priority queue. In effect, they are treated as the highest priority messages, and are executed as soon as they are picked from the network by the scheduler.

How and when to leverage priorities? As an example, consider a situation where the work consists of two types of messages: those that have only local clients and those that have remote clients (i.e., when the work is done, you have to send the result to a potentially remote chare). The latter work should have higher priority, because someone else is waiting for them, and the network latency will delay them. In general, work on the critical path of the computation can be assigned higher priority. Decisions like this are often taken after visualizing program performance via the *Projections* tool (See Chapter 3) and identifying possible bottlenecks.

Threads vs Structured Dagger: For truly reactive objects, which do not know which of their methods will be called and how many times, the generality provided by the baseline Charm++ methods is adequate and appropriate. However, if an object's life cycle is statically described, then one should use either `Structured Dagger` or a threaded entry method to code it. Threaded entry methods should only be used if `Structured Dagger` is inadequate to express the control flow. This is because a `Structured Dagger` entry method typically captures the parallel life cycle of an object, including all its remote dependencies, in a simple script, separating parallel and sequential code naturally. However, for example, if the control is deep in a function call stack, and you need a remote value, it is more convenient to use a threaded method. Threaded methods, although very efficient, are still not as efficient as the `Structured Dagger` methods: their context switching time (which is typically less than a microsecond) involves switching user-level stacks, and the need to allocate and copy/serialize the stack also adds its own overhead. Scheduling overhead for `Structured Dagger` (as well as baseline Charm++ methods) consists of a few function calls, amounting to tens to a hundred nanoseconds on current machines.

2.6 Utility of Charm++ for Future Applications

Even though Charm++ is a mature system, its signature strengths, arising from an introspective and adaptive runtime system, make it a system well-suited for addressing the challenges of the upcoming era of increasingly sophisticated applications and increasingly complex parallel machines. This is true at both the extreme scale machines, beyond the current generation of petascale computers, as well as the much smaller department-size parallel machines that are expected to be ubiquitous.

Sophisticated applications, when given a larger computer, do not increase the resolution everywhere; instead, they tend to use dynamic and adaptive refinements to best exploit the extra compute power. Sophisticated applications also use multiple modules, typically for simulating different physical aspects of the phenomena being studied. Both of these trends are likely to strengthen in the future. The dynamic load balancing capabilities, as well as the ability to support multiple modules, are critical for these applications. Yet as illustrated in this chapter, with the solid-fluid example and the SAMR example, Charm++ is well suited to express such applications with high programmer productivity.

Power and energy considerations make future machines more complex; some predictions for the future also include components failing more frequently than they do now. Again, the introspection and adaptivity, and the concomitant dynamic load balancing and fault tolerance capabilities in Charm++, help alleviate the programmer burden in dealing with such machines.

Thus, we expect the Charm++ programming model to serve the parallel applications community very well in the coming years. Of course, the runtime system itself will need improvements and modifications to cope with new hardware challenges it will surely face.

2.7   Summary

We discussed, through a series of examples, how to go about designing a Charm++ application. The stencil code illustrated how to keep the code processor-independent, and how to think about and control the grainsize of chares. Multi-physics codes such as the solid-fluid simulation, as well as molecular dynamics codes, illustrated how multiple collections of chares (Chare Arrays) can be used to cleanly express the logic of the program. Dynamic insertion/deletion and flexible indexing structures were illustrated via the structured AMR example. We also saw how to deal with task parallelism. Finally, we learned a bit more about how to use priorities and how to choose between threaded, structured-dagger and simple entry methods. Obviously, elaborating and describing the whole design process, with many additional features of Charm++ and details of their use, is beyond the scope of this book. For this, we request the reader to consult the online and other upcoming tutorials on programming with Charm++.

Acknowledgments

Work on the Charm++ system was carried out over two decades by generations of graduate students and staff members, and each has left his or her stamp on the software and the design. Many research grants over those years from U.S. agencies including the National Science Foundation, Department of Energy, National Institutes of Health and National Aeronautics and Space Administration have directly or indirectly contributed to the development of Charm++ or its applications. Some recent grants that directly funded this work are the NSF HECURA program (NSF 0833188) and the DOE FAST-OS program (DE-SC0001845). I also thank all those who helped in writing and preparing the first two chapters, including significant help from Abhinav Bhatele, especially for the many nice figures he created.

Chapter 3

Tools for Debugging and Performance Analysis

Filippo Gioachin

Hewlett-Packard Laboratories Singapore

Chee Wai Lee

Texas A&M University

Jonathan Lifflander, Yanhua Sun and Laxmikant V. Kale

Department of Computer Science, University of Illinois at Urbana-Champaign

3.1   Introduction

Appropriate software tools are essential to the effective deployment of complex HPC applications on large-scale supercomputers. Debuggers must aid developers in identifying correctness problems with their applications. They must pinpoint those problems with respect to specific regions of the code and offer hints on how they might be solved. They must even find problems outside the application's code, for example, in additional system-wide library components the application uses in a machine's software stack. Performance analysis tools in this respect are similar to debuggers. Both classes of tools share similar characteristics and face similar challenges:

1. One major goal is reducing the development and maintenance time during the lifetime of complex HPC applications. Maintenance time is often overlooked but is equally important since HPC machine software stacks change frequently. Also, input sets can exercise application functionality in unpredictable ways, particularly at very large scales.

2. They must scale to the limits of existing machines. Performance tools need to effectively gather and process information about the behavior of an application at the limits of its scaling, which is where its inefficiencies manifest. Further, bugs may reveal themselves only at specific large process counts.

3. They must minimize perturbation of the application, particularly at large scales. In the case of debuggers, the perturbation can often be sufficient to prevent the manifestation of some bugs. In the case of measurement tools, excessive perturbation can change the behavior of an application at runtime, rendering performance information ineffective.

In the rest of this chapter, we introduce two powerful tools designed for and distributed with the Charm++ Runtime System software. Section 3.2 discusses the CHARMDEBUG debugger and Section 3.3 covers the PROJECTIONS performance measurement, analysis and visualization suite. The focus of this chapter is to help the reader understand how both tools are designed around the Charm++ programming model, and explain how to use them effectively through the life of a Charm++ application with help from case studies involving real production codes.

3.2   Scalable Debugging with CharmDebug

Application development is typically divided into multiple phases: writing, debugging, testing, optimization, deployment, which are generally performed in a cycle, thus forming the application life cycle. Of these operations, code writing is one of the shortest, while debugging, testing and optimizing the software occupy most of the developers' time. This is true for both sequential and parallel programs. In fact, parallel applications suffer from all the problems a sequential application is subject to, plus those originating from the distributed nature of the algorithms. Compared to sequential programs, problems may also occur in a different location, i.e., on a different processor than the one where the problem originated. Charm++ applications suffer from similar problems as all other parallel applications. One key component that helps developers locate a problem is tightly integrating the debugger with the runtime system. The runtime system plays an important role in collecting the information from the application, and feeding it to the debugger when needed. This offers us valuable information otherwise very difficult to gather, and allows us to leverage the high scalability and asynchronous communication mechanism native to Charm++ to the advantage of the debugging infrastructure.

3.2.1 Accessing User Information

When dealing with a buggy application, one of the first questions that users ask is to view the content of their data structures. Traditionally, debuggers launch the application being debugged as a child process, and they directly inspect the content of the memory. Once loaded, they present it in a nice format to the user. This is possible also with Charm++ programs since they are written in C++. Unfortunately, it would be very challenging for normal users to identify their own data structures within the runtime system. The runtime system has many internal data structures that are used to manage the parallel entities. Furthermore, if Charm++ is compiled with multi-threading (SMP) or with BigSim emulation, this further complicates the internal representation. Other information, such as Charm++ array elements, is even more deeply embedded. While a traditional debugger can access the Charm++ internals and parse them appropriately, we wanted to separate the specifics of the implementation, which could change over time, from the functionality provided.

In the spirit of Charm++, we made use of its capability to automatically serialize data via the PUP framework. PUP, which stands for Pack-and-UnPack, is a framework which allows a user to specify a single function to coordinate the migration of an object, whether it is a chare or a plain C++ class. This framework can also serialize the data to/from the disk or network. In particular, by serializing the data to a network-connected client debugger, we can expose the useful information in an elegant way. Note that this procedure allows us to retrieve the same information regardless of the optimization applied during the compilation process, since it is the runtime itself that provides the information to the debugger.

This data is serialized with embedded information that enables the debugger to interpret the information according to the specified protocol. An example of the result obtained is displayed in Figure 3.3. Here, we can see on the left side a list of entry points that are available for breakpointing on the top, as well as a list of the messages currently enqueued on a processor on the bottom. On the right hand side, a message is fully unfolded, even though it came from a parameter marshalled entry method.

Naturally, there are many situations where the PUP routine is not available, such as in the case of user-defined C++ structures. For these, the traditional method of looking directly at the raw memory content, and interpret it, is still valid. To provide better support for all cases, CHARMDEBUG makes use of both methods.

3.2.1.1 Testing Application Behavior

Race conditions are one of the most difficult bugs to capture and correct in parallel programs. In Charm++, this translates in races between the messages sent, and the order in which they are processed. For example, imagine the situation illustrated in Figure 3.1, where processor A sends two messages α and γ to processors B and C, respectively. Processor B, after receiving β, sends a new message γ to processor C. Usually, message γ will be processed before message β. Nevertheless, there is no guarantee in Charm++ about ordering between the two messages received by processor C, and it can happen that message β is processed first. This may be a condition that was not correctly treated in the code, causing a problem at runtime. This kind of scenario is difficult to debug, as it may occur only once every thousands of executions, and always at unpredictable places.

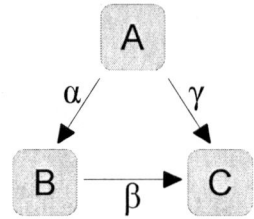

FIGURE 3.1: Example of possible race condition among three processors.

Increasing the size of the machine may trigger the problem more frequently, but at the expense of a more complex scenario to follow.

Forcing messages to be delivered out-of-order is a viable solution to hunt such a race condition. This may expose the bug very early in the application, and may not require using large machines to discover the problem. For instance, in the example above, where messages β and γ were racing, one could try imposing the delivery of β before γ even though γ appears in the queue of processor B before β. It is important that the developer understands the behavior of the application in order to guess which messages could be causing the fault. Currently, the decision of which messages to deliver provisionally is manual.

CHARMDEBUG provides a nice interface to view the data stored in the parallel application, including the messages enqueued at a certain processor. From this view, the user can easily select messages to deliver provisionally. At any time, the user can also continue inspecting the status of the system to

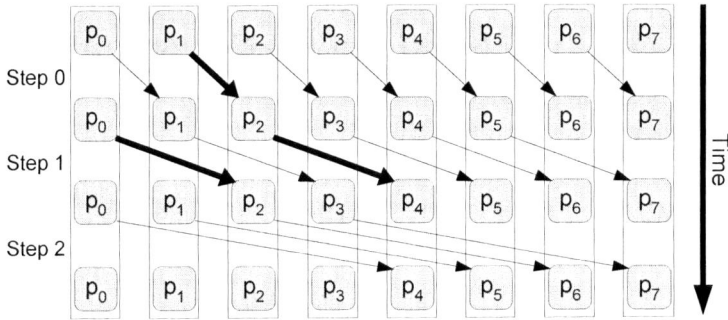

FIGURE 3.2: Parallel algorithm for prefix computation.

verify if all is proceeding as expected. In the case of a crash as a consequence of a message delivery, after having inspected the processor state, the user has the option to "undeliver" the culprit message, and continue the search in another direction. This operation is performed online and does not require the restart of the application.

To better explain the functionality of provisional delivery, we use parallel prefix as a simple example. This is a standard computation where, given an array with n elements, at the end of the computation the array will be as follows: $a_i = \sum_{k=1}^{i} a_k$. The operational flow in parallel is described in Figure 3.2. At each step i of the algorithm, processor p sends its current value to processor $p + 2^i$.

If a barrier is placed at every step of the algorithm, no problem is present. However, to increase the performance, this barrier can be relaxed, and computation can be allowed to overlap. A naïve removal of the barrier will nevertheless result in race conditions (buffering is necessary for a correct implementation). For example, the two messages highlighted in the figure destined for p_2 may arrive in any order. If the message from p_1 is processed first, the program will produce an incorrect result.

With CHARMDEBUG, the programmer can directly tweak the processing order of messages in queue, and detect the problem more easily. After the application has been started, the user can review the messages in the queue. Many of these refer to the propagation of value between elements. One of these messages has been highlighted in Figure 3.3. The user can then decide to provisionally deliver this message. The user can switch to inspect the destination object of that message (element 2 in our case), and notice that its local value has been updated to an incorrect value (i.e., not valid according to the parallel prefix algorithm). Alternatively, he or she can inspect the new messages that appear in the local queue (generated by the provisional delivery of the message), and notice that the wrong value has been sent out.

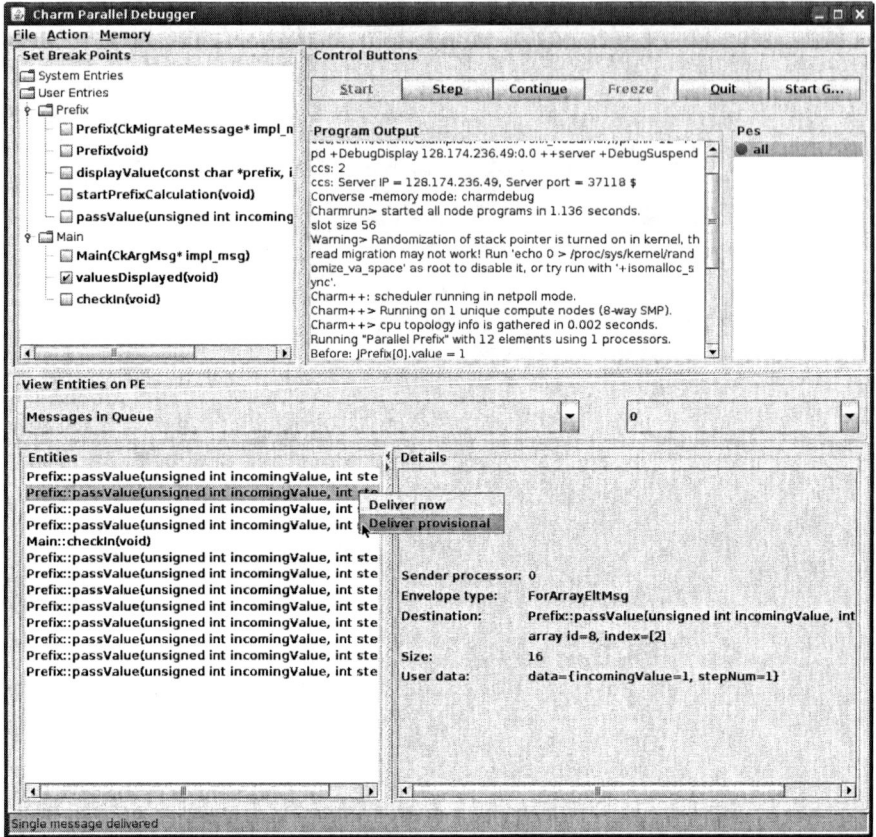

FIGURE 3.3: Screenshot of CHARMDEBUG while debugging the parallel prefix application. Multiple messages are enqueued from different steps.

If one single message delivered provisionally is not enough to understand the problem, the user can deliver multiple messages provisionally. If the computation performed is not interesting, the system can be rolled back to a previous state. This procedure can be iteratively applied until a satisfactory outcome is generated by the messages delivered.

3.2.2   Debugging Problems at Large Scale

There are many situations where an application tested on a local cluster works perfectly, but it fails when executed on a much larger machine in production mode. The reasons can be multiple, from communication mistakes to algorithm implementation errors. These bugs can be tricky to identify due to the volatility of the error: when trying to replicate the problem on a small controlled environment, the problem may not manifest any more. Generally, programmers resort to stack traces or debugging statements, like assertions or prints, to pinpoint the cause. Nevertheless, this may prove difficult since, even when running on a large machine, the problem may be intermittent, or completely disappear when inserting debugging statements that modify subtle internal timings between operations.

When deciding to attempt to debug a problem on a large number of processors, the first question is usually: will the debugger be able to handle all the processors? For CHARMDEBUG the answer is yes, provided you can run a Charm++ application on that same number of processors. The mechanism to communicate between the client GUI and the parallel application is via the CCS (Converse-Client Server) protocol, which allows clients to connect to the parallel application in a scalable manner, and internally through Charm++'s own communication layer. This guarantees high performance and scalability without additional infrastructure. In this section, we focus on the techniques that can help users to debug their parallel application in the context of Charm++. The first approach makes use of a parallel machine interactively, while the other two reduce the need for large machines during debugging.

3.2.2.1 Runtime Support for Unsupervised Execution

Even if the debugger can handle a parallel application running on thousands of processors, the programmer using the debugger may not. He can be easily overwhelmed by the amount of information presented to him. In current debuggers supporting MPI parallel application, the user still follows the traditional stepping programming model: the execution is followed on the various processors by examining the program execution line by line. While this might still work for MPI application to some extent, lockstep is unlikely to work in Charm++ given its highly asynchronous nature, where each processor may be processing completely different messages.

Instead, we leverage the Charm++ runtime system to supervise the execution on behalf of the user, and raise a notification when something unexpected happens. This enables the user to focus on what is important rather than paying attention to the whole application. The type of events that the user can be notified of can be divided in two categories: automatic and user-specified.

Automatic notifications occur when the runtime system detects that a problem has occurred, and that user intervention is required. Examples are the abnormal termination of a process due to segmentation violation or floating point exception, or the corruption of a chare by the code executed by another chare. In all these situations, the runtime system will freeze the faulty process, and allow the user to inspect the application status before shutting it down.

User-specified notifications include the traditional breakpoints, that will suspend the execution upon being hit, or the failure of an assertion specified in the code. Additionally, by exploiting the flexibility of Charm++ to process external communication interleaved with regular messages, we also allow the user to specify assertions at runtime. This is performed by writing Python

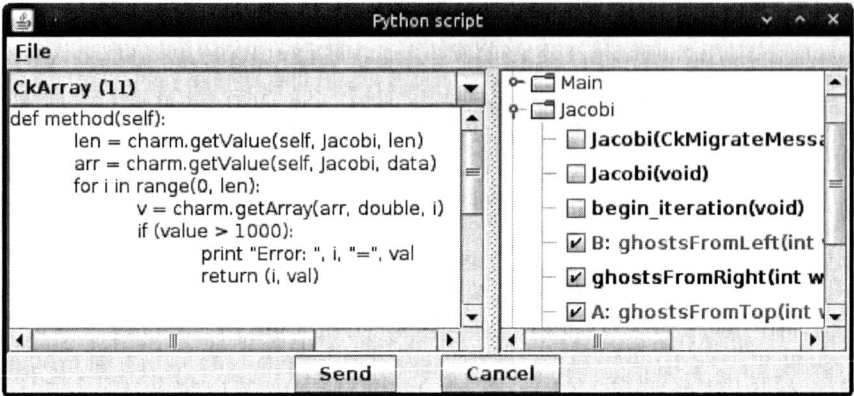

FIGURE 3.4: Screenshot of CHARMDEBUG.

scripts interactively, and uploading them to the server which can execute them after every message processed by the program, or selectively after a subset of messages [79]. This code can perform checks on the status of the system and identify problems at an early stage. The script is bound to a chare collection selected by the user, a chare array in Figure 3.4, and has access to any variable accessible to that chare. In the same figure, we can see the code on the left and that the user has selected to execute the checking code before or after certain entry methods on the right. The code in the figure performs a safety check by asserting that the values in the "data" array must be within limit.

The Python code can access variables inside the bound chare by using four helper functions. These are: *getArray*, to browse through arrays, *getValue*, to return a specific field of a data structure, *getMessage*, to return the message being delivered, and *getCast*, to perform dynamic casts between objects. All three functions return either opaque objects or simple type objects, such as int or float. If the script returns any value other than "None," the notification will be triggered, and the program suspended for further inspection.

3.2.2.2   Processor Extraction

Record-replay techniques [194, 208] are common when trying to capture non-deterministic behavior of parallel applications. In order to be effective, they must be non-intrusive, meaning that they cannot perturb the application so much that the bug disappears, and accurate, in the sense that the information recorded ought to allow a correct re-execution of the application. In Charm++, this technique may be divided into two separate algorithms: a lightweight one that records only the message order, and a comprehensive one that records the full content of each message processed. Combining these two algorithms into a three-step procedure allows the extraction of specific processors from a complex parallel application and its re-execution in a con-

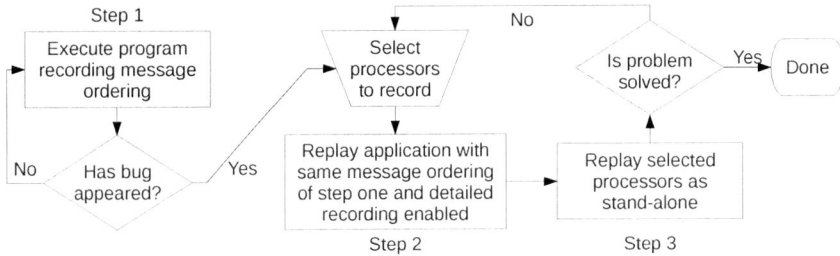

FIGURE 3.5: Flowchart of the three-step algorithm.

trolled environment. The details of this procedure are explained in detail in a previous publication [82] and illustrated in Figure 3.5.

In the first step, the entire application is repeatedly executed on the large target machine, and basic information about message ordering is recorded, until the bug manifests. The programmer then identifies a set of processors to focus on. Good candidates are processors that crash, or processors generating incorrect output. In the second step the entire application can be re-executed using the message ordering collected in step one, recording in detail the selected processors. In the third step, the detailed traces recorded in step two are used to replay a selected processor in isolation. If the problem is traced to processors that are not extracted, step two may be repeated to extract the missing processors. To perform the processor extraction, and subsequent analysis of the extracted processors, the user can switch between the command line interface and CHARMDEBUG GUI. The command line interface is especially useful when jobs have to be submitted through a batch scheduler.

Of the three steps, only the first one actually requires a large machine to be allocated. Step three requires only one processor to replay the extracted processor, while step two can be performed using BigSim [256] and emulating the large machine on a smaller one. The message ordering recorded during the first step guarantees that the execution will still be deterministic. This enables debugging even when only a few processors are available.

It is important to consider compiler optimization when debugging. While optimization may be necessary for the bug to manifest, it may hinder the usability of debugging tools such as GDB. In CHARMDEBUG, since the information recorded in the first step is independent of the particular compilation and depends only on the algorithm used, the user is allowed to switch between an optimized and a non-optimized code. In particular, the optimized version can be used in step one where timing is critical, and a non-optimized version in steps two and three. Since the message order recording scheme has a minimal impact on the application performance, using an optimized version greatly reduces the possibility of the bug disappearing.

When the bug appears only after hours of execution, maintaining all the logs in memory during step one becomes impossible. While CHARMDEBUG will automatically flush the logs periodically, this operation may disrupt the timing between messages. To solve this problem the developer can manually flush the logs to disk at appropriate times, when the disk I/O does not disrupt the timing. Since many scientific applications are iterative, the logs may be flushed when the application is transitioning between iterations. Alternatively, the developer can make use of the checkpoint/restart scheme available in Charm++ [255] to automatically checkpoint and restart from a point in time closer to the problem.

This method is by no means universal, and the programmer should understand the program being debugged to best judge if record-replay can be applied. For example, if a processor's non-determinism is due to reasons other than message ordering, such as the use of timers, record-replay may not be useful. In this case we say that the execution is not piecewise deterministic, i.e., the sequential code executed by an entry method may produce different results when executed multiple times starting from the same chare state and message content.

To demonstrate the usability of this technique, we will use CHANGA, and search for a race condition that happened during one of the computation phases. With a relatively small dataset, the bug started to appear on sixteen processors. The manifestation was intermittent, sometimes right at the beginning, sometimes after a few time steps of the application. Also, the processor in which an assertion failed kept changing from execution to execution.

Following the three-step procedure, we first executed the application and recorded the message ordering. In the execution we recorded, processor seven triggered the assertion. Subsequently, we re-executed the application in replay mode, and recorded the faulty processor. During this phase, it may be useful to test if the non-determinism has been fully captured, especially if the application uses timers or other non-deterministic routines. With the detailed trace of processor seven, we executed CHANGA sequentially under CHARMDEBUG. To track the bug we repeated the sequential execution a couple of times, each time setting a few different breakpoints. Compared to other bug-hunting techniques, this new procedure allowed for the parallel problem to be transposed into a sequential one, without compromising the timing of the application, and without allowing the problem itself to disappear.

3.2.2.3 Virtualized Debugging

Two problems can be identified that make the three-step procedure presented in the previous section not effective. The first and most important is the lack of clear identification of processors to select for extraction. This is the case when the program hangs after some time or the result is incorrect when the program terminates. In both these scenarios, no processor can be identified for extraction. Another problem is that a machine large enough to reproduce the problem may not be accessible or available.

When the user decides to proceed with a standard interactive debugging session, this may not be feasible for several reasons. Interactive sessions on large parallel machines are usually restricted to small allocations. For large allocations, batch scheduling is often required, and the application may be started at inconvenient times. This can be exacerbated by the need to execute the application repeatedly. Finally, the cost to allocate the machine may be very high, while the machine is mostly sitting idle waiting on the user to execute the next step. Instead, the user can emulate a large machine on a smaller cluster, and seamlessly debug the application as if it were running on the large configuration.

Charm++'s programming model is intrinsically bound to the concept of object virtualization, where objects are entities that "float" among processors allocated to a parallel execution. Leveraging this capability, Charm++'s BigSim project [256] aims at simulating the execution of a parallel application on large machines, even beyond the size of existing ones. By using BigSim and expanding CHARMDEBUG to access information within the emulated environment, we enabled applications to be executed on an emulated large machine while using only a smaller number of processors [81].

To demonstrate the capabilities of this technique, we present examples from a complex application, and debug it in the virtualized environment. It is not the purpose of this example to describe actual bugs that were found with this technique, but rather illustrate how the user has available all the tools compared to a normal scenario. The application we chose is CHANGA. While most of the computation is performed by Charm++ *array elements*, the application also uses Charm++ *groups* and *nodegroups*, which are bound to the number of processors involved in the simulation, for performance reasons.

The only requirement to use virtualized debugging is to build the Charm++ runtime system with support for emulation, and compile the application over the virtualized Charm++. When setting the application parameters in CHARMDEBUG, the user can input the number of real and virtual processors to be used. In case the application is launched through a batch scheduler, CHARMDEBUG automatically detects these parameters when it attaches to the running application.

Once the program has been started, and CHARMDEBUG has connected to it, the user can perform his desired debugging steps, oblivious of the fact that the system is using fewer resources internally. Figure 3.6 shows the CHANGA application loaded onto four thousand virtual processors. Underneath, we allocated only 32 processors from four local dual quad-core machines. In the bottom left part of the view, we can see all the messages that are enqueued in the selected processor (processor 3748 in the figure). Some messages have a breakpoint set (the 6th and 9th message, in medium gray), and one has actually hit the breakpoint (the 1st message). In the same message list, we can see that some messages have as a destination "TreePiece" (a Charm++ array element), while others have as destination "CkCacheManager," one of the groups mentioned earlier. One such message is further expanded in the bottom right portion of the view (the 13th message).

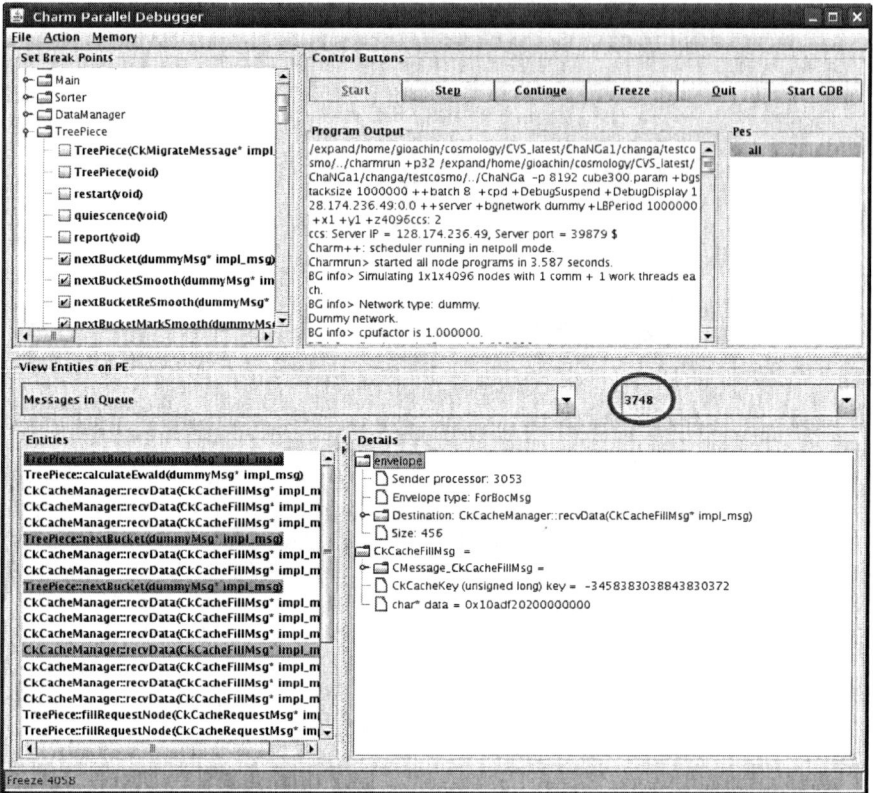

FIGURE 3.6: Screenshot of CHANGA debugged on 4096 virtual processors using 32 real processors.

There are a few warnings regarding this technique worth mentioning, in particular regarding the behavior of the application under virtualization. First, one virtual processor could corrupt the memory belonging to another one misleading the user. Second, race conditions may become more difficult to detect: by reducing the number of physical processors available, the communication latency might change such that a race condition does not appear anymore.

3.2.3   Summary

Choosing the best technique to discover the cause of a bug is not easy, and multiple techniques may lead to the same result. In this section we described the major features of CHARMDEBUG that are tailored specifically for the Charm++ programming model. These techniques leverage object-level virtualization and the message-driven programming model to help the user focus on the places bugs are more likely to be found, especially when scaling the application to larger machines. When debugging parallel applications written in Charm++, other techniques may also be used. For instance, if a bug appears on a single processor execution, a sequential debugger with appropriate breakpoints may be sufficient to discover a problem. For small scale problems, traditional methods such as print statements added to the source code may also be a viable and simple way to proceed.

3.3 Performance Visualization and Analysis via Projections

PROJECTIONS is a performance measurement, visualization and analysis tool designed for and distributed with the Charm++ runtime. It processes performance-related information associated with the key performance events in Charm++ to deliver application performance insight to analysts.

Generic performance tools like gprof [87], perfsuite [138], TAU [218], Open|Speedshop [215] and HPCToolkit [2] can be used directly on Charm++ applications to examine the time spent at the level of functions, and investigate dynamic function callpaths and callgraphs. The Charm++ runtime implementation significantly augments the performance information available to tools by exposing key events via a callback framework. These events capture information about the dynamic behavior of the runtime system when carrying out its scheduling and communication operations. The events are mapped to the Charm++ application's source code and parallel structure.

Projections was designed to take full advantage of the callbacks exposed by the runtime. It is very well suited to study Charm++ application performance at very large scales [125, 150, 213], and for a wide variety of system architectures (e.g., GPUs [231]) supported by Charm++ implementations.

The callback framework is flexible enough to be extended to support other performance tools that are adapted to use it. TAU was one such generic HPC performance tool that had successfully been adapted [23] to make use of the event callback framework.

Generally speaking, there are four distinct phases to an investigation of a Charm++ application's parallel performance:

1. **Instrumentation**, where code for measurement and collection is inserted at appropriate instrumentation points at application build time. In the case of Projections, these happen automatically through a system of event registration code fragments that are generated when the Charm++ application is built. The registration of events occurs during Charm++ initialization at runtime. These events include Charm++ communication, changes of state in the Charm++ runtime scheduler and user

code in the form of Charm++ entry methods and other model-specific parallel constructs (e.g., Structured-Dagger code [115]).

2. **Measurement and Collection**, where appropriate performance data is captured when instrumentation points are encountered at execution time. For Projections, these happen through a callback system in the Charm++ runtime in response to the presence of an instrumented trigger. The data collected is specific to each event category. Every event type registers the timestamp when they were encountered and a unique event identifier, to be resolved into a proper name identifier for visualization and analysis later. For some events, that is all that is required. Examples include the change of scheduler state to "idle" or from the "idle" state, and the handling of some newly scheduled Charm++ entry method on the local processor. In addition to the basic information, when a Charm++ entry method invokes another entry method, an outgoing message event is triggered. In this case, we also record a processor-unique tag for the outgoing event so it can be matched on the remote processor. It is important to note here that Charm++ messages can be received out-of-order and the id of the destination processor cannot be determined a priori. Finally, the invocation of a Charm++ entry method implies the receipt of an incoming message by the local runtime scheduler. Here, the id of the source processor, the source message tag as well as the size of the message, is recorded. All data is recorded into processor-local memory buffers in preparation for output.

3. **Performance Data Output** typically occurs at the end of the application's execution by flushing the performance event record buffers on each processor. The Projections measurement framework will also flush buffers whenever they are full, but this action tends to perturb the application, often making the resulting performance data output less useful. The default data output is in the form of an event trace, but Projections can also be configured to output summary profiles which are useful for basic initial performance evaluation and performance hotspot detection.

4. **Visualization and Analysis**. Here the generated performance data traces are read by a Java-based visualizer (also named "Projections") for display and for the application of various analysis options.

Projections supports two modes of performance data generation. In the first mode, performance traces are generated by linking the application with the `-tracemode projections` option. Each set of Projections trace files is accompanied by a symbol table file named <program-name>.sts. In the second mode, performance data is generated in the form of summary files, which contain per-processor profiles. This is done by linking the application with the `-tracemode summary` option. These files are more compact than traces and store summarized performance numbers corresponding to fixed-sized time

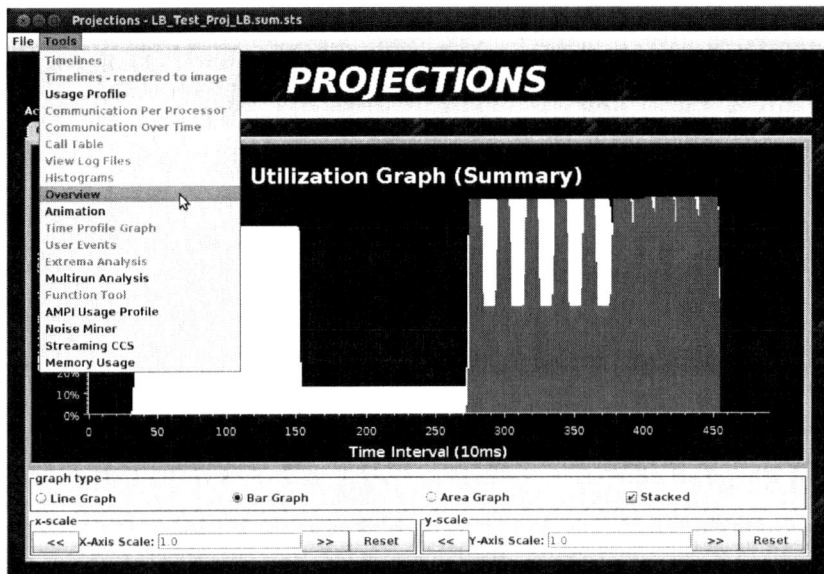

FIGURE 3.7: The main menu of the Projections visualization tool listing all its view and analysis features. In this instance, summary data had been loaded and so the inappropriate visualization features had been grayed out.

intervals over the application's execution. The set of summary files is also accompanied by its own symbol table file named <program-name>.sum.sts.

The Projections visualization tool takes summary or trace files as input for visualization and analysis. Figure 3.7 shows what the visualization tool's main menu looks like. The "Tools" option lists all available views and analysis features, graying out the ones that are not applicable to the performance data generated. It is beyond the scope of this book to describe each of those features in detail. We will instead give readers a summary of what can be accomplished through a simple example and various use cases in Sections 3.3.1 and 3.3.2. We encourage readers to peruse our online documentation[1] and tutorials on Projections for more details.

3.3.1 A Simple Projections Primer

We use a simple example based on an iterative program to demonstrate how one could use the features of the Projections visualization tool to analyze and improve program performance. This simple program creates eight Charm++ objects, each tasked to perform a different amount of work. This difference in work assigned translates directly into the amount of time spent when executed on a parallel machine. The first four of these worker objects are

---

[1]http://charm.cs.illinois.edu/manuals/html/projections/manual.html.

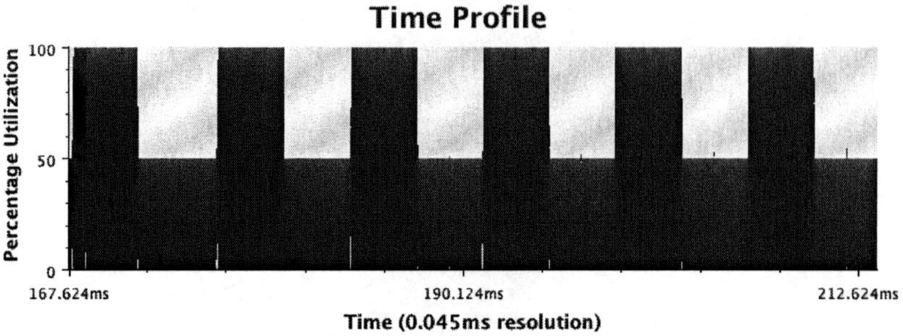

Time Profile

(a) Time profile

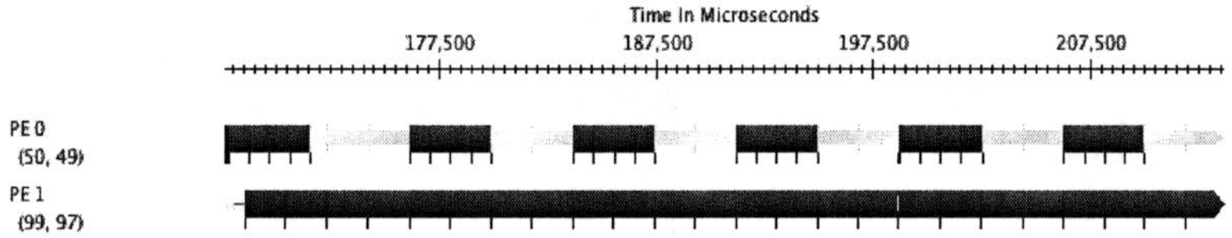

(b) Timelines

FIGURE 3.8: Running simple example on 2 processors.

assigned to processor 0 and are each tasked to perform a single unit of work while the last four are assigned to processor 1 with each tasked to perform twice the unit of work. At each iteration, each object works on its own computation and a synchronizing barrier is imposed between each iteration. Linking the program with `-tracemode projections`, as described in Section 3.3, produces performance traces which we are then able to visualize.

The first tool feature used in the typical analysis process is Time Profile view, which shows the fraction of the computation time spent in each of the Charm++ entry methods over time. This view is particularly helpful for characterizing the nature of the performance of a Charm++ program over time. In this view, the performance information represents the cumulative time taken summed across all processors. Figure 3.8(a) displays this view corresponding to the example above. It clearly shows the performance characteristics one would expect from the description of the example. The processor hosting the objects performing less work will wait for those performing more work before each iteration's synchronization point. This wait time shows up in the form of "idle time," represented by the regions of white space above the colored regions representing a Charm++ object's work. This is a synthetically engineered example of a classic load imbalance problem.

We can now turn to the more detailed timeline view to examine what the various Charm++ objects are doing. This view displays a representation of every entry method invocation, every user event, idle time and overhead time spent in the Charm++ runtime system. Users can easily select subregions,

Time Profile

(a) Time profile

(b) Timelines

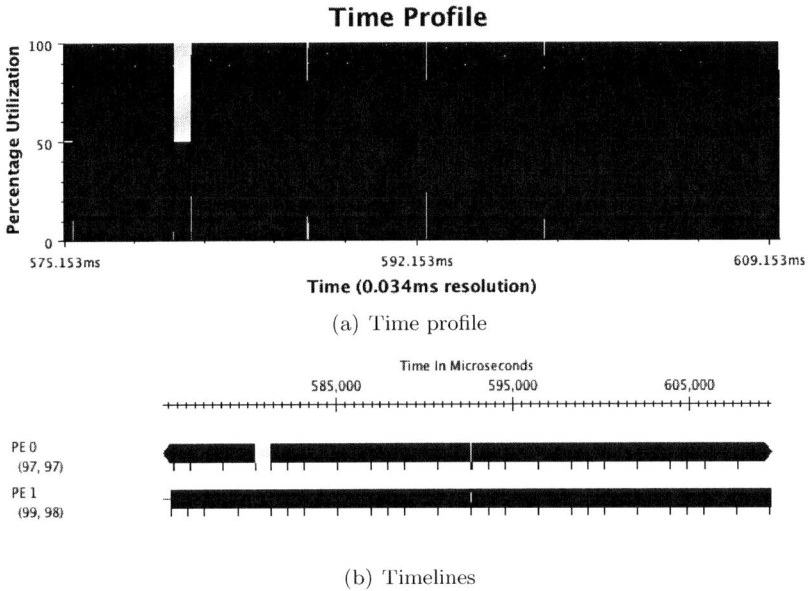

FIGURE 3.9: Running simple example with cyclic mapping on 2 processors.

zoom into or out of the selections and get information about every single entry method invocation. Figure 3.8(b) shows the equivalent timeline plot for the above example. To demonstrate how we would fix the load imbalance problem as exposed by the visualization tool, we can simply make use of various Charm++ capabilities to replace the block mapping of the objects with a cyclic mapping to allow both processors to execute the same amount of work at each iteration step. In real applications, this is often achieved by using the automated load balancing framework built into the Charm++ runtime. Figure 3.9(a) shows the results of the change, with the CPU utilization on two processors close to 100% and each processor having roughly the same amount of work. Figure 3.9(b) shows how the objects are redistributed in the timeline. It is not entirely clear why there is a small gap between the first and second iterations. It may be an artifact of the imperfect synchronization at the start of the Charm++ runtime system on behalf of the application. The fact that processor 0's activities for the first iteration are shifted slightly to the left for both Figures 3.8(b) and 3.9(b) hints at evidence supporting this hypothesis.

3.3.2 Features of Projections via Use Cases

Projections offers an extensive and rich set of tool features for studying how large-scale parallel Charm++ applications behave. As mentioned before, it would be counterproductive to attempt to list and document each feature in isolation for the purposes of this book. The following sections present real application examples and scenarios that expose some of the common use cases for the tool.

3.3.2.1   Identifying Load Imbalance

Load imbalance is a common, but important class of performance problems affecting large-scale parallel applications. The Time Profile view, as described in Section 3.3.1, provides simple but effective visual cues for identifying load imbalance. Figure 3.10(a), which shows one time step of an application, illustrates a good example. In the example, a potential load imbalance problem is visually captured by the gradual decline in overall processor utilization, represented by the height of colored bars, as the computation approaches some phase division. Note that in the case of Charm++, a phase division need not involve any explicit synchronization.

The data used for the example was generated from the production computational astronomy application CHANGA with its Barnes-Hut algorithm. CHANGA's design and implementation in Charm++ is described in detail in Chapter 6. Figure 3.10(a) shows the time profile of a CHANGA simulation with 5 million non-uniformly distributed particles over a single time step. The different colors show the execution of the various entry methods as the simulation evolves over time from left to right. The simulation was executed on 1024 Blue Gene/P processors. High utilization is achieved in the first half of

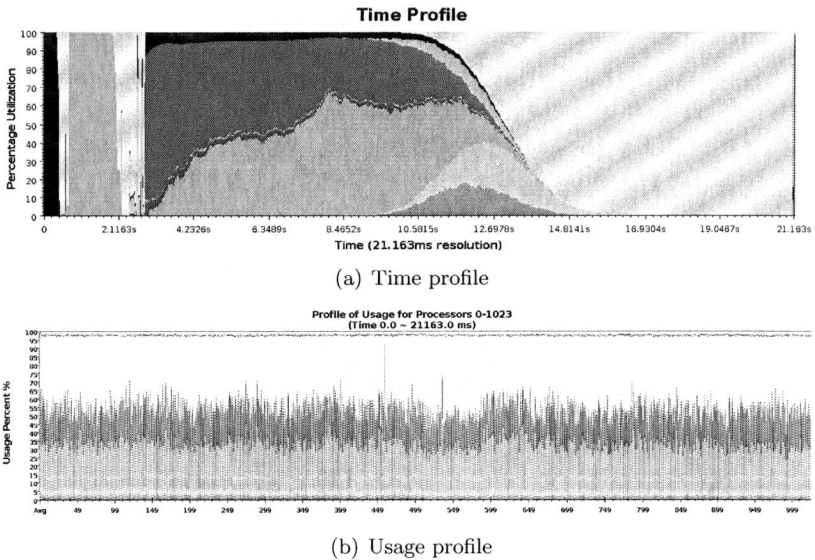

(a) Time profile

(b) Usage profile

FIGURE 3.10: Time profile and usage profile of simulating 5-million particles system in CHANGA using 1024 cores on Blue Gene/P.

the time step. However, the long tail observed in the second half of the time step is indicative of load imbalance.

To better understand this problem in this non-trivial case, we can turn to other tool features in Projections. The usage profile tool displays the break-down in entry method utilization for each processor averaged over the duration of a CHANGA time step. Figure 3.10(b) shows the usage profile display of the same execution duration corresponding to the time profile in Figure 3.10(a). A big variance of CPU utilization among the processors can be clearly seen in the usage profile. More importantly, one can observe several processors with significantly higher overall utilization than others. This supports the hypothesis that load imbalance exists. The reason is simple. It takes only one heavily overloaded processor prior to a phase division or synchronization to delay every other processor.

It is easy to see that when the number of processors involved is large, it is difficult to manually locate the heavily-loaded processors using the usage profile feature. For this purpose, Projections supports several features oriented towards scalability of visualization and analysis. To address this specific problem, Projections provides the *Extrema tool* as a means to find the usage profiles of individual processors. The Extrema tool sorts, filters and displays processor usage profiles by some desired criterion. The user selects the desired criterion from a set of pre-defined ones. An example of a commonly used pre-defined criterion is the sorting, filtering and display of usage profiles based on total CPU utilization. Figure 3.11(a) shows all the processors sorted and displayed from left to right in increasing order by total CPU utilization. Projections is flexible enough to display only a specified number of extrema processor usage profiles. Users are not forced to see information for all processors, but only the most significant by criteria.

The Extrema tool in Projections is also designed to be linked to the time-line view. By selecting any usage profile bar displayed by the Extrema tool, the user can choose to quickly pull up the timeline for that specific processor. We demonstrate this by using the Extrema tool to add the five most heavy-loaded processors to the timeline view, resulting in Figure 3.11(b). We are able to then use the timeline view to examine how the objects behave on these processors as well as how objects from other processors interact with them. The timeline view facilitates quick examinations of object-to-object interactions across processors via simple mouse-clicks, automatically loading more processor timelines as necessary. This enables users to discover which processors and objects were involved in interaction-dependency chains, derived backward from the most delayed events in the timeline. For example, in Figure 3.11(b), one might begin by examining the communication events that trigger the execution of the entry methods at the tail-end of delay processor 458.

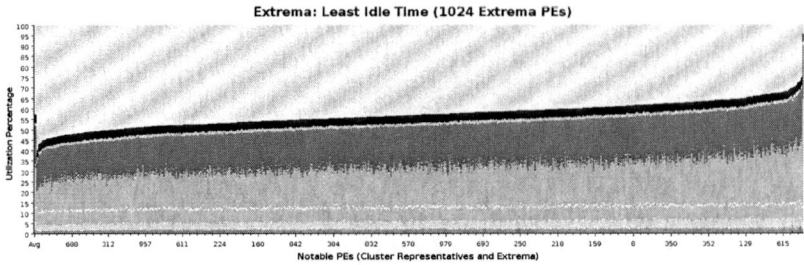

(a) Least idle processors

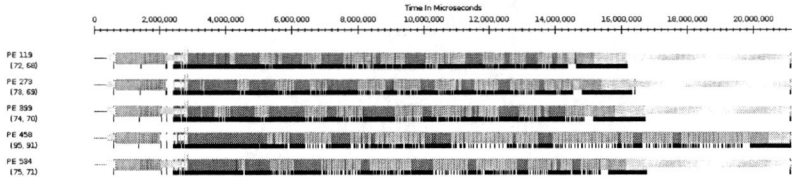

(b) Timeline of least idle processors

FIGURE 3.11: Extrema and timeline view of simulating a 5-million particles system in CHANGA using 1024 cores on Blue Gene/P (**see Color Plate 1**).

3.3.2.2   Grainsize Issues

Chapter 2 defines and describes the concept of *grainsize* in Charm++. From a performance analysis perspective, the grainsize of an entry method in Charm++ can be summarized to represent the time taken to execute the work encapsulated in any single invocation of the entry method. Insight into the scalability characteristics of an application may be gained by studying its execution's grainsize distribution.

One challenge in Charm++ programming is that the programmer must decompose the problem into tasks while balancing two requirements. The decomposition must create enough parallelism to fully utilize all resources, while keeping the overhead of task creation and scheduling to a minimum. If there are too many tasks that take a long time (i.e., the application is coarse-grained), then any task dependency chains in an application that involve the participation of those lengthy tasks will limit the scalability and performance of the application. If there are too many tasks that take too little time (i.e., the application is fine-grained), then the scheduling overhead of the Charm++ runtime may become significant. As a general rule-of-thumb, however, it is often better to over-decompose the problem domain and allow the runtime to dynamically adapt to the mix of tasks than to be conservative with grainsize.

To help visualize the distribution of tasks of various grainsizes, Projections provides the *Histogram tool*. The number of instances of entry method invocations are counted and tallied against their recorded execution duration.

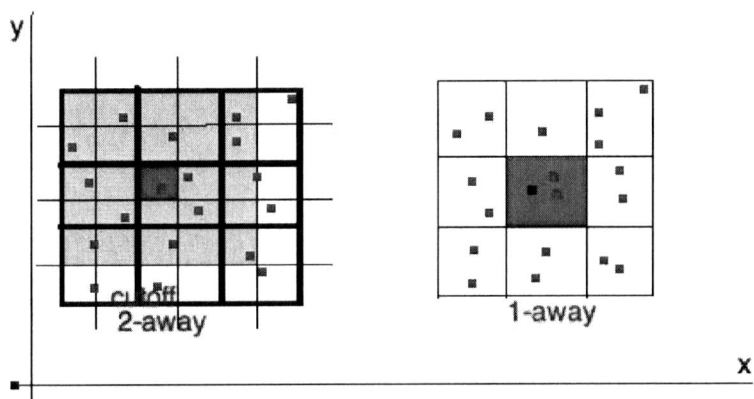

FIGURE 3.12: 1-away and 2-away decomposition in NAMD.

These counts are organized according to fixed-duration bins and displayed color coded based on the application's entry methods. By analyzing the histogram, parallel programmers can decide how to adjust the grainsize in the code. For example, a programmer might split the long tasks into smaller ones to avoid the load imbalance caused by the long tasks.

In the molecular dynamics application NAMD, described in Chapter 4, we use both spatial decomposition and force decomposition to increase parallelism. However, for some systems, when the NAMD simulation is scaled to large number of cores, the total utilization can suffer.

Figure 3.12 illustrates a high level abstract view of two decomposition methods used in NAMD. In 1-away decomposition, the computation tasks encapsulating interactions between atoms are defined between neighboring patches. In 2-away decomposition, the same spatial domains are sub-divided and finer-grained computation tasks are defined for each of the expanded set of smaller interacting patches. Note that the validity of any domain decomposition method is subject to the science and mathematics governing the simulation. Observe that in the case of the 2-away decomposition, interactions still need to be computed based on the same spatial distance between atoms, with some corrections. By applying and comparing these two decompositions methods to the same molecular simulation, one can observe significant performance differences. For instance, when simulating ApoA1—a 92K-atom molecular system using NAMD on 1024 cores of a Cray XE6 Machine—the time taken for one iterative step of the simulation with 1-away decomposition is 3.9*ms*, while it is 3.1*ms* for 2-away decomposition. Figure 3.13 shows the time profile displayed by Projections for four time steps of an ApoA1 simulation on 1024 cores with 1-away domain decomposition. Figure 3.14 shows the ApoA1 simulation with the same execution configuration and over an approximately similar execution phase, but with 2-away domain decomposition.

Time Profile

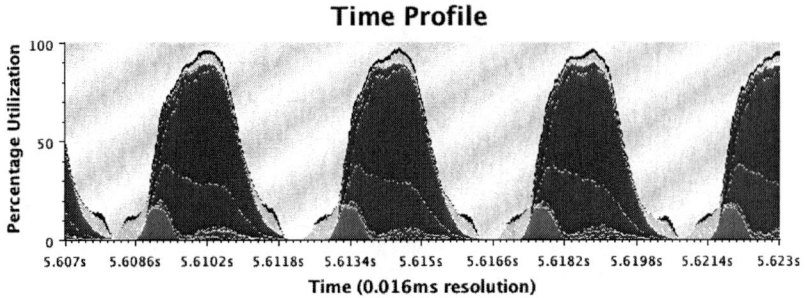

FIGURE 3.13: Time profile of simulating ApoA1 on 1024 cores with 1-away decomposition.

The time profiles' display patterns hint at some form of load imbalance. Indeed, one should be able to apply the same techniques described in Section 3.3.2.1 to eventually figure out that the most time-consuming task dependency chains tended to include tasks that take more time than is typical.

The histogram tool in Projections, however, provides a more appropriate mechanism to help users reveal the primary reasons for the performance difference. Figure 3.15(a) illustrates the grainsize distribution of parallel tasks when using 1-away decomposition. We can see how the grainsize of the tasks vary from $0ms$ to $1.0ms$, in $10\mu s$ intervals, along with the entry method type and the number of those tasks. We notice a tri-modal distribution of grainsize, with the longest mode centered around $0.7ms$, with a significant number of methods taking more than $1ms$, as indicated by the last bar (which counts methods of duration $1ms$ or above). With the 2-away decomposition for the same molecular simulation, Figure 3.15(b) shows that the grainsize is now within a much tighter range of execution duration from $0ms$ to $0.28ms$ and

Time Profile

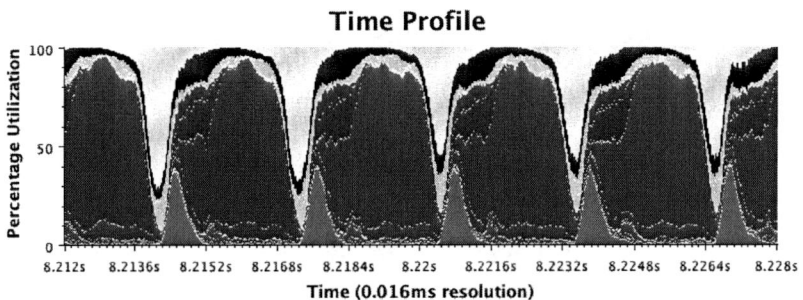

FIGURE 3.14: Time profile of simulating ApoA1 on 1024 cores with 2-away XY decomposition (**see Color Plate 2**).

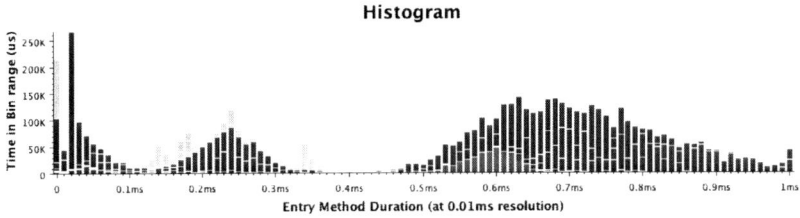

(a) 1-away

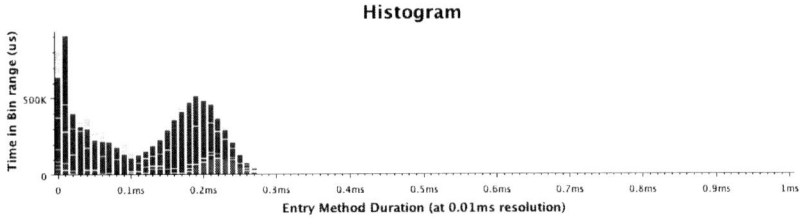

(b) 2-away

FIGURE 3.15: Histogram of simulating ApoA1 on 1024 cores with 1-away and 2-away XY decomposition.

with many more tasks. This validates the hypothesis that we can achieve performance gains for this particular simulation by switching from a 1-away domain decomposition to a 2-away scheme. Going back to Figure 3.14, one can observe that the average CPU utilization over time has increased and each step takes a shorter amount of time. This can clearly be seen by the presence of five time steps over the same execution duration instead of the four displayed in Figure 3.13.

3.3.2.3   Memory Usage

As we scale to larger systems, the amount of memory per core is expected to decrease. Some applications already utilize much of the available memory. The ability to understand and optimize the application's memory usage over time is important. In particular, it is important to understand how the scheduling of entry methods by the Charm++ runtime scheduler can cause memory to increase or decrease over time.

Dense LU factorization in Charm++ is an example that can demonstrate problems resulting from memory constraints. Here, the Charm++ tasks representing trailing updates for matrix-to-matrix multiply operations arrive on a processor, but may have to be delayed in order to execute other tasks that are on the critical path and require higher priority processing. However, delaying these trailing updates will cause memory usage on that processor to

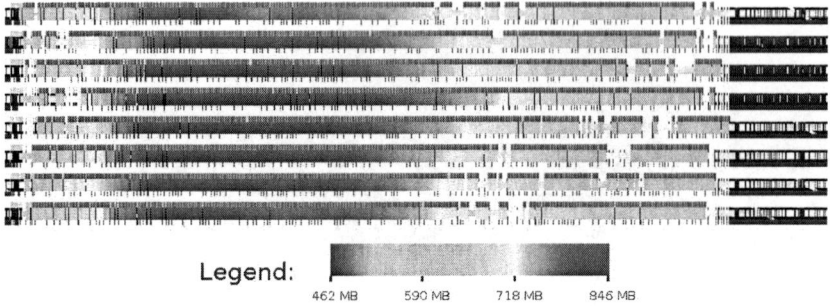

Legend:

462 MB   590 MB   718 MB   846 MB

FIGURE 3.16: Memory usage as visualized by Projections via a modified timeline. Entry methods are colored according to their memory usage. A legend shows the color scale and the data values (**see Color Plate 3**).

increase. Figure 3.16 shows Projections coloring the entry methods by their memory usage with a corresponding legend for its color key. Users can use this feature to quickly examine memory hot spots in the application and to make the necessary changes to the scheduling of tasks to alleviate those hot spots.

3.3.3   Advanced Features for Scalable Performance Analysis

It is important to remember that the effective use of Projections as a tool requires it to be scalable itself. Scalability issues ultimately come from too much performance information generated.

3.3.3.1   Application Execution Duration

For Projections, traces produced by linking the application with the `-tracemode projections` option, the file size for each processor will grow relative to the duration of the application's execution time if it is not controlled. This has two possible and orthogonal effects: large unwieldy files that will affect the performance of the visualization and analysis toolset; and/or the frequent flushing of trace buffers each time Projections' buffer space runs out. The latter can cause severe application performance perturbations.

Charm++ application developers have several ways of tackling problems associated with tracing. If file size is the only issue, the `+gz-trace` runtime option will produce compressed traces. Please note that compressed traces will continue to impact the visualization tool's performance the same way uncompressed traces do, but will consume less space on disk. The more common approach is to restrict the period during which performance information is recorded into the trace buffers. This is achieved via the API calls `traceBegin()` and `traceEnd()` that can be inserted into the Charm++ code at appropriate points. It is also possible to start the application with tracing turned off using the `+traceoff` runtime option. The performance analysis of applications like NAMD usually revolves around several hundred iterative time steps selected from an appropriate execution time period.

3.3.3.2   Number of Processors

Of the two dimensions of application scaling this is probably the more important, given that Projections will generate a trace or summary data per processor. Thankfully, the visualization system is equipped to handle partial trace log data and still present a consistent and relatively coherent overall picture of application performance. If one is forced to generate large trace logs for each processor, it is still possible to analyze the data by manually removing a subset of the logs. One could choose to analyze blocks of processors (e.g., 0 to 511) separately or strides of processors (e.g., every 5 processors).

A more sophisticated but experimental method [150] exists for automatically selecting a subset of important performance logs from a large set of processors. The `+extrema` runtime option activates a K-means clustering algorithm at the end of the application's execution which determines the mapping of processors to a computed set of behavioral equivalence classes. The algorithm picks a set of processors that are exemplars of each equivalence class and a set of processors that represent outliers (or extrema) with respect to each equivalence class. The union of the sets of outliers forms the set of all outliers in the application. The former represent probable baseline behavior in each class while the latter represent possible candidates for poor or unexpected performance behavior that analysts ought to examine more closely. The `+numClusters` runtime option allows developers to control the number of performance classes that are discovered. As Projections currently employs a cluster seeding scheme that does not guarantee non-empty sets, this option establishes the upper bound in the number of non-empty equivalence classes that will be discovered.

3.3.4   Summary

The Projections framework provides a rich and powerful feature set for the collection, presentation and exploration of performance characteristics in Charm++ applications. It is a highly scalable tool, which has been used to analyze performance in runs with over 300,000 cores, based on detailed traces. In this chapter, we presented some of these features through the use of case studies. Extensive documentation for the tool framework is available online and many more use cases can be found in our technical publications which we will highlight in Section 3.4.

3.4   Conclusions

When building parallel applications using the Charm++ programming model, developers have to deal with the various aspects of code writing, debugging and performance optimization. Having specific tools capable of understanding the programming model and providing useful information when needed is a key benefit to improve productivity and to generate high quality and efficient software.

CHARMDEBUG and Projections are two important tools, which we used in the debugging, tuning and optimization of many of our flagship scientific applications written in the Charm++ programming model. In particular, Projections has been used in the optimization process for achieving high scalability in NAMD [167, 140], CHANGA [108] and OPENATOM [25]. It has also been used to study the adaptive overlap of computation and communication in multi-physics simulations [110].

Projections is used extensively to study the performance characteristics and effects of many Charm++ features and techniques when applied to a wide variety of large-scale parallel algorithms, codes and machines. For example, Sun et al. [232] demonstrated the use of Projections in comparing the performance differences of the Charm++ communication layer implemented directly on the user Generic Network Interface (uGNI) against the MPI implementation on Cray's Gemini interconnect. The latest studies involving Projections include the observation of performance effects due to load balancing when seeking good trade-off decisions between energy consumption and performance.

Acknowledgments

The authors would like to thank Isaac Dooley, Orion Lawlor and Gengbin Zheng, for their significant contributions to the tools infrastructure. Their contributions have facilitated the development of the work described in this chapter. We also thank the many generations of students who have worked on the CHARMDEBUG and Projections code base. Our deepest appreciation goes to our users for their patience and feedback when using our tools.

The work presented in this chapter was supported by these grants: NIH grant P41-RR05969-04, NSF grants ITR PHY-0205413 and DMR 0121695, DOE grant B523819 and the NASA AISR program. This research used resources of the Argonne Leadership Computing Facility at Argonne National Laboratory, which is supported by the Office of Science of the U.S. Department of Energy under contract DE-AC02-06CH11357. This work used the Extreme Science and Engineering Discovery Environment (XSEDE), which is supported by National Science Foundation grant number OCI-1053575 (project allocations TG-ASC050039N and TG-ASC050040N).

Chapter 4

Scalable Molecular Dynamics with NAMD

James C. Phillips and Klaus Schulten

Beckman Institute, University of Illinois at Urbana-Champaign

Abhinav Bhatele

Center for Applied Scientific Computing, Lawrence Livermore National Laboratory

Chao Mei

Intel Corporation

Yanhua Sun, Eric J. Bohm and Laxmikant V. Kale

Department of Computer Science, University of Illinois at Urbana-Champaign

4.1   Introduction

The NAMD software, used by tens of thousands of scientists, is focused on the simulation of the molecular dynamics of biological systems, with the primary thrust on all-atoms simulation methods using empirical force fields, and with a *femtosecond* time step resolution. Since biological systems of interest are of fixed size, efficient simulation of long time scales requires the application of fine-grained parallelization techniques so that systems of interest can be simulated in reasonable time. This need to improve the time to solution for the simulation of fixed sized systems drives the emphasis on "strong scaling" performance optimization that engendered this collaboration between physical and computer scientists.

Performance improvements motivated by NAMD inspire abstractions, optimized implementations and robust infrastructure in Charm++, and complementary improvements to Charm++ enable the implementation of new features in NAMD. The collaborative and synergistic development underlying the NAMD project (started by principle investigators, Klaus Schulten, Laxmikant V. Kale and Robert Skeel, in 1992) has contributed to many important achievements in molecular modeling, parallel computing and numerical algorithms. As recognized in the 2012 IEEE Computer Society Sidney Fernbach Award, jointly awarded to Kale and Schulten, NAMD has been an important contribution to the scientific community.

In this chapter, we will discuss the motivation for biomolecular simulation (Section 4.2), parallelization techniques for molecular dynamics (Section 4.3), the parallel design of NAMD (Section 4.4), its application to ever larger scale simulations (Section 4.5), overall performance (Section 4.6) and elaborate upon a few of NAMD's applications (Section 4.7).

4.2   Need for Biomolecular Simulations

The form and function of all living things originate at the molecular level. Genetic information in nucleic acids encodes the sequence of amino acids for proteins, which once assembled by the ribosome, fold into the specific three-dimensional structures that enable their function in the cell.

Cellular proteins can be isolated, purified and grown into crystals, from which X-ray diffraction can be used to determine the positions of the protein atoms with great accuracy. Larger aggregates, such as the ribosome, can be studied through cryo-electron microscopy, and the resulting coarse images combined with high-resolution crystal structures to obtain atomic resolution for the complete aggregate. While these experimentally determined structures alone are of great utility in explaining and suggesting mechanisms for the observed chemical and mechanical behavior of biomolecular aggregates, they represent only static and average structures. The detailed atomic motions that lead to function cannot be observed experimentally.

Physics-based simulations step in where experiment leaves off, allowing the study of biomolecular function in full atomic detail. Although atomic interactions are governed by quantum mechanics, the energies found in biological systems are sufficiently low that chemical bonds are only formed or broken in the reaction centers of catalytic proteins. As a result, atomic interactions in biomolecules can be represented via simple classical potentials for electrostatics, van der Waals and bonded interactions. While this simplification greatly reduces the computational demands of simulations, many orders of magnitude are required to extend size from atoms to cells and time from femtoseconds to seconds.

Current state-of-the-art simulations may follow millions of atoms for mere microseconds, often employing additional techniques to enhance sampling. Smaller, longer-time simulations may follow the entire protein folding process. Larger simulations allow the study of aggregates such as the ribosome, which builds proteins in all cells and is a common target of antibiotics, the chromatophore, which is the basic photosynthetic unit of plants and the protein capsids of viruses, which bind to and penetrate the cell membrane to enable the infection process. A recent example is the ground-breaking study, published in *Nature* [250], that determined the structure of the HIV capsid, based on a NAMD simulation with 64 million atoms.

4.3   Parallel Molecular Dynamics

Molecular dynamics (MD) simulations follow molecular systems ranging from a few thousand to millions of atoms for tens of nanoseconds to microseconds. When doing these simulations sequentially, the time period to be simulated is broken down into a large number of time steps of 1 or 2 femtoseconds each. At each time step, forces on each atom (electrostatic, van der Waals and bonded) due to all other atoms are calculated and the new positions and velocities are determined. The atoms are moved to their new positions and the process repeats.

Parallelizing a MD simulation is challenging because of the relatively small number of atoms and large number of time steps involved. Traditionally, three different methods have been used to parallelize MD simulations: atom decomposition, spatial decomposition and force decomposition [199, 200]. Atom decomposition involves distributing the atoms in the MD simulation among the processors evenly. Each processor is responsible for the force calculations for its atoms. Spatial decomposition is similar except that the physical simulation space is divided up spatially to assign atoms to different processors. Force decomposition, on the other hand, involves creating a matrix of force calculations to be performed for pairs of atoms and assigning responsibility for the calculation of a part of the matrix to each processor.

Atom and force decomposition have a high communication-to-computation ratio asymptotically whereas spatial decomposition suffers from load imbalance problems. NAMD pioneered the hybrid decomposition scheme in which the processors holding the atoms and those calculating the forces are decoupled [121]. The parallelization scheme is a hybrid between spatial and force decomposition. Atoms in the simulation box are divided spatially into smaller boxes and assigned to some processors. The force calculations for a pair of sub-divided boxes are assigned to an arbitrary processor which can be different from the ones holding the two boxes. The scheme is described in detail in the next section. Similar parallelization schemes have been used in other recent MD packages such as Desmond [26] and Blue Matter [65], and in the scheme proposed by Snir [223].

4.4   NAMD's Parallel Design

NAMD is one of the first scientific applications to use Charm++ as the underlying runtime system. Over the last decade, NAMD development has fueled Charm++ research and instigated new features and capabilities in the runtime. In turn, NAMD has benefited from features such as dynamic load balancing and section multicasts that are a part of Charm++.

NAMD has co-developed with Charm++, and served as a confirmation of the utility of some of the features of Charm++, such as message-driven execution and load balancing. An early version of NAMD, which was a precursor to the current version, was written in the mid 1990s. Two different versions were maintained for some time: one was in PVM, and the other in Charm, the C-based precursor to Charm++. The modularity benefits of Charm++ started becoming clear in comparing these variants, especially as they were developed further. The PVM version needed to have a message-driven loop explicitly in its code. Messages belonging to different modules were dealt with in this one loop, and it had to keep track of the progress of different modules, and even different instances of them (such as multiple sub-domains, computation objects and long-range force calculation objects). In contrast, in the Charm++ version, the objects belonging to different modules were cleanly separated. They naturally allowed adaptive overlap of communication with the computation across modules, yet required no breaching of abstraction boundaries.

At the same time, the separation of objects from processors, which was the hallmark of the original NAMD design, was very naturally supported by Charm++'s object model. The separation of the collection of objects for doing force calculations from the objects housing the atoms also allowed us to (and required us to) do explicit load balancing. With such load balancing, Charm++ was able to exploit an arbitrary number of processors within a reasonable range. In other words, there were no restrictions on the number of processors having to be a cube or even a product of three integers, as was typically required in molecular dynamics applications based on spatial decomposition.

4.4.1   Force Calculations

Molecular dynamics simulations involve calculating forces on each atom (electrostatic, van der Waals and bonded) due to all other atoms. A naïve pairwise calculation of non-bonded forces between all pairs of atoms has a time complexity of $\mathcal{O}(N^2)$. In order to reduce this complexity to $\mathcal{O}(N \log N)$, forces are calculated explicitly only within a cutoff radius, r_c. Beyond this distance the forces are calculated by extrapolating the charge densities of all atoms to a charge grid and using the particle-mesh Ewald (PME) [42] method.

NAMD uses a hybrid decomposition scheme that separates the distribution of data (atoms) from the distribution of work (force calculations). The implementation of the hybrid decomposition scheme and independent calculation of different types of forces is facilitated by the ability to create multiple sets of chares in Charm++ that can be mapped independently to the processors. The simulation box is divided spatially into smaller boxes called "patches" which collectively form one set of chares (see Figure 4.1 which shows a simplified two-dimensional simulation space). The number of patches can be less than the number of processors, in which case the patches are assigned to a subset of the processors. Force calculations between a pair of patches are assigned to chares from another set called the compute objects, or just "computes." There are three different types of computes—1) bonded computes that calcu-

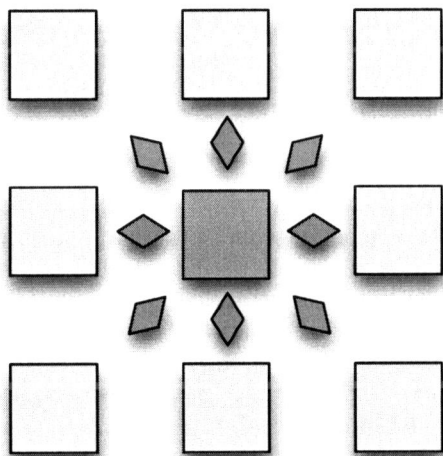

FIGURE 4.1: Hybrid decomposition in NAMD (the square objects are patches and the diamond objects are non-bonded computes).

late the forces due to bonds, 2) non-bonded computes that are responsible for calculating short-range non-bonded forces and 3) PME computes, responsible for calculating long-range electrostatic forces.

Each non-bonded compute is responsible for the force calculations between a pair of patches (or a single patch in case of interactions between atoms within a patch). Hence, each patch sends its atoms to several computes (nine in case of the 2D decomposition shown in Figure 4.1, twenty-seven in case of a 3D decomposition) whereas each compute receives atoms from two patches. The sending of atoms from a patch to all its computes is done via a section multicast that creates a "spanning" tree between the processor holding the patch (root) and the processors holding the associated computes. Forces from all computes to a given patch are also sent back along this tree.

Three different chare arrays are used for the PME computation which uses a two-dimensional pencil decomposition of the charge grid for parallelization: PMEZPencils, PMEYPencils and PMEXPencils. The patches communicate with the PMEZPencils at the beginning and completion of each PME phase. There are several line FFTs within the pencils in each direction and transpose operations between pencils in different directions. Since this phase has a relatively small amount of computation and intensive communication (due to the transposes), it is often done every four time steps instead of every time step.

4.4.2   Load Balancing

The presence of different kinds of chares, patches, bonded computes, non-bonded computes, and three types of PME computes makes load balancing a formidable task. However, the load balancing framework in Charm++ is designed to handle multiple chare arrays in the application. The load balancing framework is measurement-based and relies on the *principle of persistence* of load. This principle assumes that the load distribution in the recent past is a reasonable indicator of that in the near future. The runtime instruments all the chares in the application for their execution times and also records the communication graph between them. This graph is made available to the load balancing framework to make migration decisions. Applications can plug in specific strategies that exploit application-specific knowledge for a better load balance.

In the case of NAMD, all chares—patches, bonded computes, non-bonded computes and PME computes—are instrumented for their execution time. The total load on each processor is the sum of the execution times of all objects that reside on it. The loads on each processor in the previous time steps are used to make decisions about migrating the chares for better balance. Only the non-bonded computes, which account for a significant fraction of the execution time, are made migratable. The rest of the computes are assigned statically during program start-up but their loads are considered when balancing the migratable objects.

Load balancing in NAMD is periodic. Before the load balancer is invoked, a few time steps are instrumented and that information is used for balancing load for future time steps. The first time that load balancing is performed, the algorithm reassigns all migratable objects. Subsequent calls perform a refinement-based load balancing that minimizes migrations by preserving the previous assignments as much as possible. The load balancing strategy is a greedy heuristic-based algorithm that creates a max heap of objects and min heap of processors (based on their execution times) and maps objects iteratively starting with the heaviest ones to the least loaded processors.

The Charm++ runtime provides a detailed communication graph of the chares involved in an application to the load balancer. The balancing algorithm can use the communication information to minimize communication as well as migration. This information is also used for optimizing the communication on the underlying network topology in the case of torus machines such as the IBM Blue Gene platforms [22]. An interconnect topology aware mapping of the patches and computes in NAMD can optimize communication on the network and minimize network congestion.

4.5 Enabling Large Simulations

The unprecedented growth in the size of parallel machines and the requirements of computational scientists to simulate molecular systems with tens to hundreds of millions of atoms have put the scaling performance of NAMD to test and resulted in significant improvements to the software to enable such use cases.

4.5.1 Hierarchical Load Balancing

Traditionally (before 2010), the load balancing strategies in NAMD were executed serially by collecting the instrumented data (loads and communication graph) on one processor. This becomes infeasible when running a large molecular system or on a large number of processors or with a large number of chares. Collecting the entire communicating graph on one processor and then sending migration decisions out from it leads to a serialization bottleneck in messaging. Storing this information in the memory of one node also becomes infeasible. Finally, the serial load balancing algorithm running on one processor can take a very long time to execute while all other processors are idle, waiting for the decisions. These factors motivated the use of a hierarchical load balancing scheme in NAMD [252].

NAMD uses the hierarchical load balancing support available in Charm++ [251]. In this scheme, the processors are divided into independent groups that are arranged in a hierarchy forming a tree. The tree can have any number of levels and an arbitrary number of children per node. Every node and

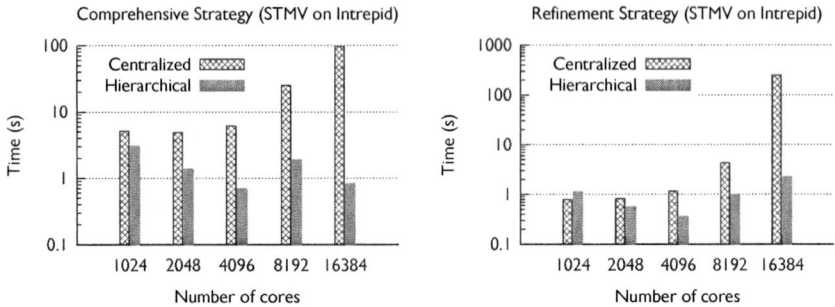

FIGURE 4.2: Improvements in load balancing time from using hierarchical load balancing in NAMD on IBM Blue Gene/P (Intrepid).

its immediate children at the next level form an autonomous group. Within each group, a root node or group leader performs load balancing serially for all processors within its group. At higher levels, group leaders represent the entire sub-tree below them. Load information is first exchanged bottom up and then load balancing is done in a top-down fashion. Within each group, existing load balancing strategies in NAMD such as comprehensive and refinement can be invoked.

Use of hierarchical load balancers leads to significant reductions in the memory consumption in NAMD and more importantly huge savings in the execution time of the load balancing strategies. Figure 4.2 shows the reduction in the time spent in the comprehensive and refinement load balancing strategies simulating the Satellite Tobacco Mosaic Virus (STMV). On 16,384 cores of IBM Blue Gene/P, the time spent in load balancing is reduced by more than 100 times! This improvement is attained while retaining a high quality of load balance achieved, so application performance is almost as good as that with centralized load balancers.

4.5.2   SMP Optimizations

Multicore nodes in parallel machines have motivated the design and implementation of a multi-threaded SMP runtime mode in Charm++ [168]. In this mode, each Charm++ processing element (PE) runs as a thread as opposed to an OS process in the non-SMP runtime mode. All threads (i.e., Charm++ PEs) belonging to the same OS process form a Charm++ "node." The nature of a single memory address space shared by Charm++ PEs on a "node" enables several optimization opportunities.

Reduce Memory Footprint: Read-only data structures or immutable ones (only written once) can be shared among Charm++ PEs on a "node." Exploiting this opportunity can also lead to other benefits such as better cache

No. of nodes	140	560	2240	4480	8960	17920
No. of cores	1680	6720	26880	53760	107520	215040
non-SMP (MB)	838.09	698.33	798.14	987.37	1331.84	1760.86
SMP (MB)	280.57	141.83	122.41	126.03	131.84	157.76
Reduction factor	2.99	4.92	6.52	7.83	10.10	11.16

TABLE 4.1: Comparison of average memory footprint between SMP and non-SMP during simulation (12 cores per node).

performance. In NAMD, optimizations are done to share certain information such as the molecule object that contains static physical attributes of atoms and map objects that track the distribution of patch and compute objects. Table 4.1 shows the comparison of average memory usage per core when running NAMD in non-SMP and SMP modes on the Jaguar machine at Oak Ridge National Laboratory, demonstrating the effectiveness of reducing the memory consumption using SMP mode. In addition, we also observed much better cache performance directly related with the memory footprint reduction [167].

Improve Intra-node Communication: Charm++ PEs on the same "node" can also exploit the use of shared memory address space to improve the performance of communication. Instead of making a copy of the message when performing intra-node communication in the non-SMP mode, the Charm++ runtime simply transfers the memory pointer of the message in the SMP mode. This optimization is transparent to the application and is embedded in the runtime. Therefore, NAMD automatically enjoys the benefits from the improved intra-node communication. In addition, considering faster communication within a "node," if a message is sent to multiple PEs on the same "node" from a PE on a remote "node," we can optimize this communication scenario by just sending one inter-node message and then forwarding this message to destination PEs within a node. In this way, the expensive inter-node communication is replaced with the more efficient intra-node one. We refer to this as node-aware communication optimization, and it is exploited as much as possible in communication idioms such as the general multicast/broadcast in the Charm++ runtime and NAMD-specific multicast operations [167].

Exploit More Fine-grained Parallelism: Several computation functions in the NAMD PME phase have been observed to execute on a few cores on each node with idle neighboring cores. To improve NAMD's performance during this phase, the fine-grained parallelism inherent in those computation functions needs to be exploited and distributed among the idle neighboring cores. OpenMP provides a language directive based approach to realizing this. Using OpenMP threads in conjunction with Charm++ PEs on the same cores is not straightforward because they are not aware of each other. We have developed a "CkLoop" library for the SMP mode of Charm++ to use the Charm++ PEs to mimic the responsibilities of OpenMP threads. Reference [231] shows the performance benefits from using this "CkLoop" library for the PME computation.

Reduce Usage of System Resources: In the SMP node, significantly fewer OS processes are created in the parallel application. This implies that the usage of system resources which depends on the number of processes also reduces. The benefit of such reduction is exemplified by the decrease in the job launch time. When running NAMD on 224,076 cores of Jaguar where Charm++ is built to rely on MPI to perform communication, mpirun takes about 6 minutes to launch the job in the non-SMP mode where each core is hosting a MPI rank. In comparison, in the SMP mode, each node is hosting a MPI rank instead, covering 12 cores. As a result, mpirun then only takes about 1 minute to launch the NAMD job.

4.5.3   Optimizing Fine-Grained Communication in NAMD

As described earlier, most of the time in the PME phase in NAMD is spent in communication. When scaling NAMD to large number of nodes this communication in PME scales poorly and eventually becomes the major performance bottleneck. Therefore, it is crucial to optimize this communication pattern. Below, a few techniques for optimizing the PME communication are discussed.

Increasing PME message priority. In NAMD, various types of messages play different roles in driving the program execution. For example, the non-bonded function in the compute object is performed when the messages containing atoms from its two patches arrive. Similarly, the arrival of PME messages drives the FFT computation. When different types of messages are queued to be sent or to be executed, the selection of messages to process first can potentially affect the overall performance. When the PME phase becomes the performance bottleneck, it is highly desirable to process the PME messages as soon as possible. In order to do this, we assign PME messages with higher priority than other messages. Two techniques are applied to implement this idea. On the sender side, messages with high priority are processed first. Only after these messages are injected into the network, the other messages get a chance to be processed. On the receiver side, instead of processing messages in a first-come-first-serve (FCFS) order, incoming messages are queued in the order of priority. Therefore, the computation driven by messages with high priority is performed first. With these two techniques, the delay of processing PME messages is minimized which improves the overall performance and scaling significantly.

Persistent communication for FFT. For most applications in Charm++, when messages are sent on the sender, the memory to store the message on the destination is usually unknown. Only when the message arrives at the receiver, the corresponding memory is allocated for it. However, in scientific applications, we have observed that there is "persistent" communication, which means that the communication partners and message sizes for one transaction do not change across time steps. There are two possible benefits of exploiting this persistent communication. First, we can save the time to allocate/free memory on the receiver. The other benefit is to better exploit the underlying network if it supports remote direct memory access (RDMA). Without using persistent communication, three steps are required to send a message on the RDMA network. First, a small control message including sender's information is sent to the receiver. Based on the information in this small message, the receiver performs a RDMA "get" transaction to transfer the real data. An ack message is sent back to the sender to notify the completion of the data transfer. Compared with this, using persistent communication, the sender has the information of the receiver so that a direct RDMA "put" can be issued to transfer the data. Hence, the small control message is avoided. In NAMD, we implemented the communication in PME with persistent messages. A 10% performance improvement is observed when running a 100-million-atom simulation on the Titan machine at ORNL.

4.5.4 Parallel Input/Output

As we started to simulate very large molecular systems with tens of millions of atoms using NAMD on hundreds of thousands of processors, we found that the input/output (I/O) in NAMD, i.e., loading molecular data at start-up and outputting atoms' trajectory data to the file system, became a major roadblock. Since existing parallel I/O libraries such as HDF, netCDF, etc. do not handle NAMD file formats, we chose to implement parallel I/O natively in NAMD. One main advantage, enabled by the asynchronous message-driven programming model of Charm++, is that we can then optimize for writing trajectory data frame-by-frame, overlapping with the computation on other processors during the simulation.

Traditionally, NAMD loads and processes all molecular data on a single core before broadcasting the data to all other cores. Although this approach is adequate for moderately large molecular systems, it does not scale to molecular systems with several million atoms due to the inherent sequential execution. For example, it requires nearly an hour and about 40 GB of memory to initialize a 100-million-atom STMV simulation on a single core of an Intel Xeon (1.87 GHz) processor. To address this issue, we first developed a compression scheme by extracting "signatures" of atoms from the input data to represent the common characteristics that are shared by a set of atoms. Together with atoms' "signature" input, a binary file containing the information of each atom, constructed from the original input file, is fed into the native parallel input scheme described as follows.

A small number of processors are designated as input and output processors. Considering P "input" processors, one of them first reads the signature file and then broadcasts this to all other input processors. P is usually smaller

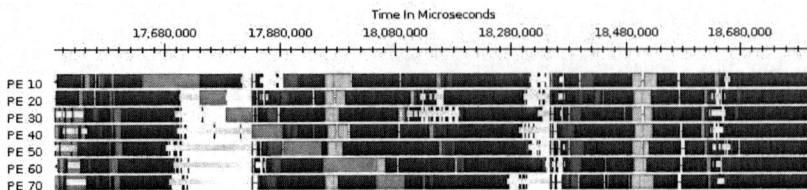

FIGURE 4.3: Parallel Output Timeline of a 2.8-million-atom Ribosome Simulation on Jaguar (Cray XT5).

than the total number of processors and can be automatically tuned to optimize for the memory footprint and performance. After the initial broadcast, each of these P processors loads $1/P$ of the total atoms starting from independent positions in the binary file. Then they shuffle atoms with neighbor input processors according to molecular grouping attributes for later spatial decomposition. Comparing with the sequential input scheme, for the 100-million-atom STMV simulation, this parallel scheme with 600 input processors on Jaguar (Cray XT5 at ORNL) completes the initialization in 12.37 seconds with an average memory consumption of 0.19 GB on each input processor, a $\sim 300\times$ reduction in time and a $\sim 200\times$ reduction in memory footprint!

We faced similar performance and memory footprint challenges in the output of trajectory files, but with an additional one posed by maintaining fast execution time per step under tens of milliseconds in case of frequent output. Similar to the parallel input scheme, with a tunable number of "output" processors, each output processor is responsible for the trajectory output of a subset of atoms. Furthermore, we have implemented a flexible token-based output scheme in which only those output processors that have a token could write to the file system in order to handle I/O contention on different parallel file systems. Reaping benefits from Charm++, the file output on one processor can potentially overlap with useful computation on other processors as clearly illustrated by Figure 4.3 showing the tracing of a single-token-based output scheme. In the figure, the output activity represented by the light gray bars clearly overlaps with useful computation (dark gray) on other cores spanning multiple time steps.

4.6   Scaling Performance

NAMD is run on a variety of supercomputer platforms at national supercomputing centers in the U.S. and elsewhere. It has demonstrated good strong and weak scalability for several benchmarks on different platforms. The platforms vary from small memory and low frequency processors like the IBM

System	No. of atoms	r_c (Å)	Simulation box	Time step (fs)
ApoA1	92224	12	$108.86 \times 108.86 \times 77.76$	1
F1-ATPase	327506	12	$178.30 \times 131.54 \times 132.36$	1
STMV	1066628	12	$216.83 \times 216.83 \times 216.83$	1
Ribosome	2820530	12	$264.02 \times 332.36 \times 309.04$	1
100M STMV	106662800	12	$1084 \times 1084 \times 867$	1

TABLE 4.2: Simulation parameters for molecular systems used for benchmarking NAMD.

Blue Gene machines to fast processors like the Cray XT5 and XK6. The size of molecular systems ranges from benchmarks as small as IAPP with 5570 atoms to an STMV system with 100 million atoms. Table 4.2 lists the various molecular systems (and their simulation details) that were used for obtaining the performance numbers presented here.

Performance for various molecular systems: Figure 4.4 shows the execution time per step for five molecular systems running on an IBM Blue Gene/P (BG/P). ApoA1 and F1-ATPase scale well up to 8192 cores while the bigger systems of STMV and Ribosome scale up to 16,384 cores. The 100-million-atom system has demonstrated scalability up to almost the entire machine at Argonne (Intrepid, 163,840 cores). The simulation rate for ApoA1 at 16,384 cores of BG/P is 47 nanoseconds per day (ns/day) or 1.84 ms per time step. The simulation rate for the 100M STMV system at 131,072 cores is 0.78 ns/day or 111.1 ms per time step.

Performance on various machines: Figure 4.5 presents the execution times for the 1- and 100-million-atom STMV systems on different machines: Blue Gene/P, Ranger and Cray XK6. The 1-million system scales well on the

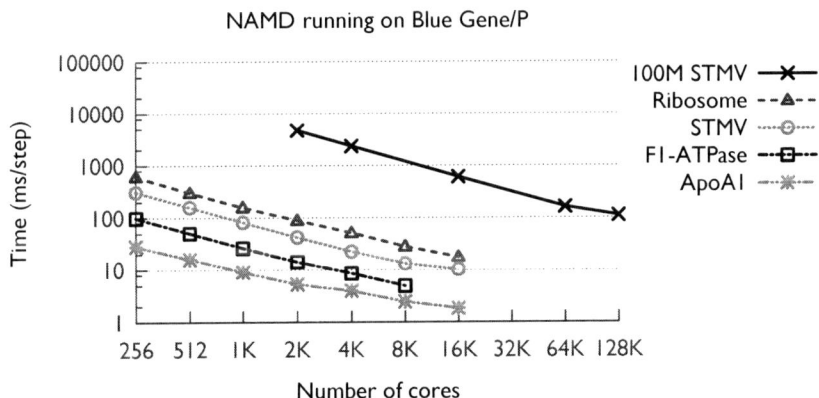

FIGURE 4.4: Performance of NAMD on IBM Blue Gene/P (Intrepid).

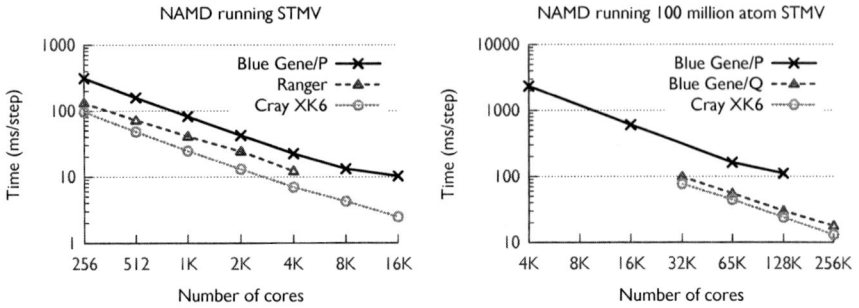

FIGURE 4.5: NAMD performance on several machines.

different platforms (left plot). The best performance is achieved on the Cray XK6 with much faster cores and a high-speed interconnect compared to the BG/P. At 16,384 cores, the execution time is 2.5 ms per step (simulation rate of 34.6 ns/day). The benchmarking results for the much bigger 100-million-atom STMV running on the Blue Gene/P, Blue Gene/Q and Cray XK6 are shown in the right plot. NAMD demonstrates excellent scalability for this molecular system on the Blue Gene/Q and XK6 with good performance at as many as $262,144$ cores.

Extreme strong scaling of NAMD: One needs to execute a billion time steps to simulate $1\mu s$ in the life of a biomolecular system! Further, a particular system being studied has a fixed size, i.e., a fixed number of atoms. So, in order to do the simulations faster, one needs to carry out a single time step as fast as possible. What are the current limits of such strong scaling? Recently, we were able to simulate a time step in about 500 μs. This corresponds to a simulation rate of 170 ns/day. Given the amount of communication, coordination and critical-path-bound work one has to do in each time step, such numbers are impressive, and are a testament to NAMD's innate performance orientation, as well as the design of the machines themselves.

Table 4.3 shows the best performance we have achieved for ApoA1 with PME every 4 steps on IBM Blue Gene/Q and Cray XC30. On Blue Gene/Q there are fewer than two atoms per core and we are really pushing the scaling

Machines	Nodes	Cores	Atoms per Core	Time (ms/step)
Blue Gene/Q	4096	65536	1.4	0.683
Cray XC30	512	8192	11.0	0.526

TABLE 4.3: Time step of running ApoA1 with PME every 4 steps on Blue Gene/Q and Cray XC30.

limit using extremely fine-grained decomposition. On Cray XC30, in part due to new Cray Aries interconnect, better performance is obtained on fewer nodes. In both cases, the time per step is below one millisecond, which brings us closer to the goal of simulating longer time scales in the life of biomolecules.

4.7   Simulations Enabled by NAMD

NAMD is distributed free of charge as both source code and convenient pre-compiled binaries by the NIH Center for Macromolecular Modeling and Bioinformatics at the University of Illinois. NAMD is a popular program with over 50,000 registered users in the past decade, over 16,000 of whom have downloaded multiple versions of the program. NAMD has been cited in over 5000 publications and is one of the most used programs at NSF-funded supercomputer centers. NAMD development is driven by the projects of the NIH Center, examples of which are presented below (also see Figure 4.6).

In the year 1999, NAMD enabled the study of the photosynthetic purple membrane of *Halobacterium salinarium*, simulating a hexagonal unit cell

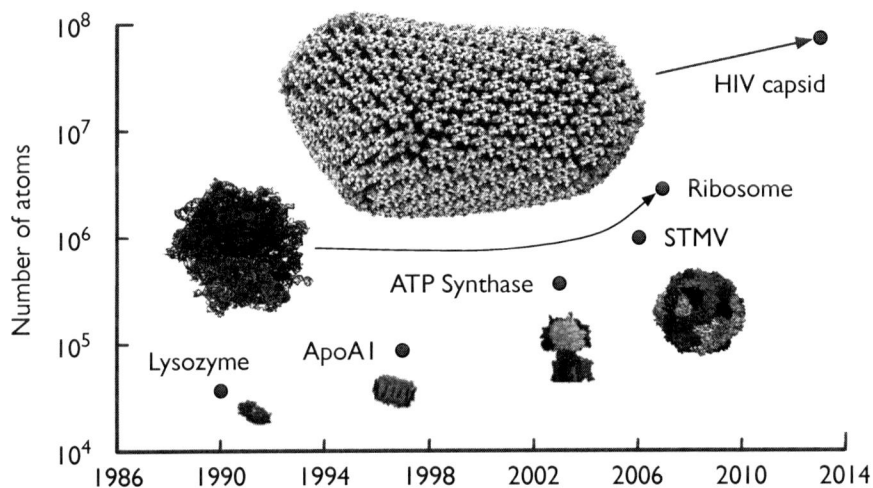

FIGURE 4.6: The size of biomolecular systems that can be studied using all-atom molecular dynamics simulations has steadily increased from that of a Lysozyme (40,000 atoms) in the 1990s to the F_1F_0-ATP synthase and STMV capsid at the turn of the century, and now 64 million atoms as in the HIV capsid model shown above. Atom counts include aqueous solvent, not shown (**see Color Plate 4**).

containing 23,700 atoms distributed over protein, lipid membrane, ion and water components [17]. The simulations were used to study how proteins in the membrane capture light energy to pump protons across the membrane. The difference in proton concentration thus established provides energy source that another membrane protein, adenosine triphosphate (ATP) synthase, stores as chemical bonds in ATP for transport and utilization in other processes in the cell [233]. Beginning in 2001, much larger (327,000 atoms) NAMD simulations of ATP synthase were used to study this process. In 2002, a NAMD simulation of a membrane containing aquaporin proteins revealed the mechanism by which aquaporin water channels permit water and other neutral molecules to cross the membrane while preventing the passage of protons and charged molecules [234]. Thus, aquaporins solve the important problem of maintaining a proton gradient to drive ATP synthesis while allowing water to pass.

In 2005, NAMD enabled the first all-atom molecular dynamics study of a complete virus particle [67]. Satellite Tobacco Mosaic Virus (STMV), a small and well studied plant virus, was simulated as a complex of over one million atoms of protein, nucleic acid and water. By studying the stability of the complete virion and its isolated components, the simulations illustrated that previous speculation that STMV assembly was mostly capsid protein-driven was likely incorrect, and that instead the virus's genetic payload recruits capsid proteins into a shell around itself.

By 2012, this initial work on virus simulation had matured such that the human HIV virus capsid could be studied, enabled by NAMD simulations on petascale machines such as the "Blue Waters," a Cray XE6 at Illinois. Initial 10-million-atom simulations of a cylindrical HIV assembly have now been extended to a 64-million-atom simulation of the full HIV capsid [250]. Similarly, the earlier studies of photosynthesis have progressed to models of a complete photosynthetic unit, a pseudo-organelle called the chromatophore consisting of several hundred proteins embedded in a spherical lipid membrane [216]. Planned simulations of the chromatophore will exceed 100 million atoms.

4.8   Summary

NAMD is the first science application to use Charm++ as its underlying parallel framework. The Charm++ and NAMD collaboration has come a long way and has benefited both programs immensely. Performance improvements motivated by NAMD have inspired abstractions and optimized implementations and robust infrastructure in Charm++, and complementary improvements to Charm++ have enabled the implementation of new features in NAMD. NAMD is one of the best scaling parallel molecular dynamics packages and portable to almost any architecture by virtue of using the Charm++ runtime.

NAMD is installed at major supercomputing centers in the U.S. and around the world and is used by many research groups for their simulations. The study of the influenza virus (A/H1N1) and the HIV capsid are testimony to the impact of NAMD in the field of biophysics and drug design.

Acknowledgments

The many contributors to the development of the NAMD software made this project a success, including: Bilge Acun, Ilya Balabin, Milind Bhandarkar, Abhinav Bhatele, Eric Bohm, Robert Brunner, Floris Buelens, Christophe Chipot, Jordi Cohen, Jeffrey Comer, Andrew Dalke, Surjit B. Dixit, Giacomo Fiorin, Peter Freddolino, Paul Grayson, Justin Gullingsrud, Attila Gursoy, David Hardy, Chris Harrison, Jérôme Hénin, Bill Humphrey, David Hurwitz, Barry Isralewitz, Sergei Izrailev, Nikhil Jain, Neal Krawetz, Sameer Kumar, David Kunzman, Jonathan Lai, Chee Wai Lee, Charles Matthews, Ryan McGreevy, Chao Mei, Esteban Meneses, Mark Nelson, Ferenc Ötvös, Jim Phillips, Osman Sarood, Ari Shinozaki, John Stone, Johan Strumpfer, Yanhua Sun, David Tanner, Kirby Vandivort, Krishnan Varadarajan, Yi Wang, David Wells, Gengbin Zheng and Fangqiang Zhu.

This work was supported in part by a grant from the National Institutes of Health NIH 9P41GM104601 "Center for Macromolecular Modeling and Bioinformatics." This research is part of the Blue Waters sustained-petascale computing project, which is supported by the National Science Foundation (award number OCI 07-25070) and the state of Illinois. Blue Waters is a joint effort of the University of Illinois at Urbana-Champaign and its National Center for Supercomputing Applications. This work was performed under the auspices of the U.S. Department of Energy by Lawrence Livermore National Laboratory under Contract DE-AC52-07NA27344 (LLNL-BOOK-608433).

This research used resources of the Argonne Leadership Computing Facility at Argonne National Laboratory, which is supported by the Office of Science of the U.S. Department of Energy under contract DE-AC02-06CH11357. This research also used resources of the Oak Ridge Leadership Facility at the Oak Ridge National Laboratory, which is supported by the Office of Science of the U.S. Department of Energy under Contract No. DE-AC05-00OR22725. This work used the Extreme Science and Engineering Discovery Environment (XSEDE), which is supported by National Science Foundation grant number OCI-1053575 (project allocations TG-ASC050039N and TG-ASC050040N).

Chapter 5

OpenAtom: Ab initio Molecular Dynamics for Petascale Platforms

Glenn J. Martyna

Physical Sciences Division, IBM Research

Eric J. Bohm, Ramprasad Venkataraman and Laxmikant V. Kale

Department of Computer Science, University of Illinois at Urbana-Champaign

Abhinav Bhatele

Center for Applied Scientific Computing, Lawrence Livermore National Laboratory

5.1   Introduction

OPENATOM is parallel simulation software for studying atomic and molecular systems based on quantum chemical principles. In contrast to classical computational molecular dynamics which is based on Newtonian mechanics, OPENATOM uses the Car-Parrinello *ab initio* Molecular Dynamics (CPAIMD) approach. This allows it to study complex atomic and electronic physics in semiconductor, metallic, biological and other molecular systems. The application has been designed to expose maximal parallelism via small grains of data and computation. The resulting implementation atop Charm++ is highly scalable, and has exhibited portable performance across three generations of the IBM Blue Gene family, apart from other supercomputing platforms.

Instead of using an empirical force function, the CPAIMD algorithm computes the forces acting on each atom as the summation of multiple terms derived from plane-wave density functional theory. Unlike traditional bulk-synchronous parallelization that simply decomposes the data, OPENATOM exploits the underlying mathematics via a seamless mix of both data and functional decompositions. This results in greater expressed parallelism, and several overlapping phases of computation combined with a longer critical path of dependent computations.

Such a design is enabled, and greatly facilitated, by the Charm++ tenet of *parallel program design and decomposition using units that are natural to the application domain.* Instead of dividing data into as many pieces as processors, OPENATOM simply decomposes the data *and* the computation across a number of chare objects. The type or number of these pieces is not limited by the number of processors. Rather, they depend on the CPAIMD algorithm and the desired grainsize. For example, an electronic state is a unit of data that is natural to the CPAIMD algorithm and is one of the types of objects in the application.

We attempt, in this chapter, to further expand on such an approach to designing successful, scalable parallel programs (Section 5.3). We preface the description of our parallel design with a discussion of the underlying physics (Section 5.2). This includes a description of the computational algorithm, as well as the time and space complexities of each portion of the computation. The success of such a design approach is substantiated with performance results in Section 5.5. Like several other successful Charm++ applications, OPENATOM has also inspired abstractions, libraries and other features that have made it back into the Charm++ parallel programming ecosystem. Section 5.4 briefly describes some of these features. We finally conclude with a few scientific studies that have used OPENATOM and the work planned for the future.

5.2 Car-Parrinello Molecular Dynamics

Car-Parrinello *ab initio* Molecular Dynamics (CPAIMD) [32, 203] is a key computational technique employed in the study of structure and dynamics of atomistic systems of great interest throughout science and technology (S&T). The number of citations to the original research paper has grown exponentially and the method's use has spread from the physical sciences of chemistry, biology, geology and physics into the materials science and engineering disciplines. CPAIMD has indeed become an essential and ubiquitous tool for the investigation of the properties of matter of all types.

The power of the CPAIMD method lies in the novel combination of increasingly accurate electronic structure (ES) methods with increasingly efficient molecular dynamics (MD) techniques in such a way that they can be simulated/solved on the largest parallel High Performance Computing (HPC) platforms in existence today. Combining ES and MD allows the study of highly complex atomistic systems that involve changes in chemical bonding patterns or simply non-standard bonding under both equilibrium and non-equilibrium conditions. Unlike stand-alone *ab initio* methods where the atoms are typically fixed or may move along only an energy minimized pathway, CPAIMD allows the atoms to evolve naturally under the influence of Newton's equation of motion, molecular dynamics, with forces derived from ES theory. In this way, the effects of pressure, temperature and field gradients on systems with complex electronic structure, for example, can be discerned and the properties of liquids and amorphous materials that do not have a single identifiable representative structure can be illuminated. CPAIMD can be coupled to advanced sampling MD techniques to increase the time scales that can be accessed, and with path integral methods to determine nuclear quantum effects such as tunneling to increase the range of validity of the technique.

The CPAIMD method has been successfully applied in geophysics to describe the behavior of the cores of gas giant planets [33], in chemistry to understand the fundamental principles of aqueous acids and bases [163], in physics to study the properties of metal-insulator transitions [217], in engineering to study the behavior of devices and in materials science to study novel materials such as complex oxides [219]. This and other seminal work has had an important impact across S&T, leading to new scientific insight and engineering applications.

At present CPAIMD is limited to systems dominated by their ground state ES properties; it assumes the Born-Oppenheimer approximation wherein the nuclei evolve on a potential energy surface formed by the electronic ground state energy and nuclear-nuclear Coulombic interactions. The accuracy of ES methods intrinsic to the CPAIMD technique, which are necessarily approximate, is not currently sufficient to treat some critical systems, such as diradicals, and systems dominated by dispersion interactions, such as biological membranes, with tractable computational efficiency [39]. The CPAIMD method is often applied using a plane-wave basis set to describe the electronic states within the Gradient Corrected Local Density Approximation (GG-LDA or GGA) [18, 149, 195] to Density Functional Theory (DFT) [103, 135]. Research is underway to improve all aspects of the CPAIMD technique so as to increase accuracy, computational efficiency and applicability.

One of the important factors that has led to the wide adoption of the CPAIMD method is the availability of highly (parallel) scalable, user-friendly HPC software. Some of the major plane-wave based DFT packages include CPMD, Quantum Espresso, AbInit, QBOX and OpenAtom. CPMD, QBOX and OpenAtom have superior parallel scaling; AbInit, CPMD and Quantum Espresso have large user bases while OpenAtom is a Charm++ based experimental package designed and used primarily for CS based parallel HPC software and scientific physics-based methodological development in addition to materials research. CPMD, Quantum Espresso and OpenAtom have fairly open user licenses. All code bases have produced important application studies highly relevant to S&T.

5.2.1 Density Functional Theory, KS Density Functional Theory and the Local Density Approximation

Density Functional Theory states that the ground state energy of an electronic system can be expressed, exactly, as the minimum of a functional of the electron density [103, 175],

$$E[n(\mathbf{r})] \quad = \quad \int d\mathbf{r} n(\mathbf{r}) v^{ext}(\mathbf{r};\mathbf{R}) + F[n(\mathbf{r})] - \mu \left[\int d\mathbf{r} n(r) - n_e \right] \quad (5.1)$$

$$\frac{\delta E[n(r)]}{\delta n(\mathbf{r})} \quad = \quad = 0 \quad\quad\quad\quad\quad\quad\quad\quad\quad\quad (5.2)$$

Here, the e-nuclear interaction potential is $v_{ext}(\mathbf{r}; R)$, and the unknown functional, $F[n(\mathbf{r})]$, can be expressed as the sum of physically intuitive terms:

$$F[n(\mathbf{r})] \quad = \quad E_H[n(\mathbf{r})] + T[n(\mathbf{r})] + E_{xc}[n(\mathbf{r})]$$

$$E_H[n(\mathbf{r})] \quad = \quad \int d\mathbf{r} \int d\mathbf{r}' \frac{n(\mathbf{r})n(\mathbf{r}')}{|\,\mathbf{r}-\mathbf{r}'\,|}$$

The Hartree energy (E_H) is the interaction of classical charge distributions, the electronic quantum kinetic energy is T and the "exchange-correlation" functional is E_{xc}, which accounts for Fermi-statistics ("exchange") and other many body quantum effects ("correlation"). The Lagrangian multiplier μ insures the density represents the correct number of electrons n_e and is physically the chemical potential of the system. In general, the exchange-correlation functional may be separately divided into an exchange part (which is known exactly in certain limits) and a correlation part. Again, when minimized, $E[n(\mathbf{r})] = E_0$.

In order to allow the development of good approximate functionals, Kohn and Sham decomposed the electron density into a sum over a set of orthonormal electronic states [135],

$$n(\mathbf{r}) \quad = \quad \sum_s \Psi_s^2(\mathbf{r}) \tag{5.3}$$

to yield

$$E[n(\mathbf{r})] \quad = \quad \int d\mathbf{r} n(\mathbf{r}) v^{ext}(\mathbf{r}; \mathbf{R}) + F[n(\mathbf{r})] - \sum_{ss'} \lambda_{ss'}$$

$$\left[\int d\mathbf{r} \Psi_s^*(\mathbf{r}) \Psi_{s'}(\mathbf{r}) - 2\delta_{ss'} \right]$$

$$F[n(\mathbf{r})] \quad = \quad E_H[n(\mathbf{r})] + T_S[n(\mathbf{r})] + E_{xc,KS}[n(\mathbf{r})]$$

$$T_S[n(\mathbf{r})] \quad = \quad -\frac{\hbar^2}{2m} \sum_s \int d\mathbf{r} \Psi_s(\mathbf{r}) \nabla^2 \Psi_s(\mathbf{r}) \tag{5.4}$$

where each electronic state is doubly occupied, consistent with the Pauli-exclusion principle (1-spin down and 1-spin up electron occupy each state). We restrict ourselves to the spin-paired electron case here. A set of Lagrange multipliers $\lambda_{ss'}$ assures the normalization of the states. The exchange correlation functional is now relative to the non-interaction system and hence noted E_{xc}^{KS}; this distinction shall be dropped below. A widely used approximate functional is termed the Gradient Corrected Local Density Approximation (GG-LDA or GGA) to DFT and is written as [18, 149, 195]

$$E_{xc}[n(\mathbf{r})] = \int d\mathbf{r} \epsilon_{xc}(n(\mathbf{r}), \nabla n(\mathbf{r})) n(\mathbf{r}) \tag{5.5}$$

We restrict our discussion in this chapter to the GG-LDA approximation to KS-DFT and, hereafter, simply refer to the technique as "DFT" to preserve simplicity.

In the discussion below, the nuclear-nuclear interaction,

$$\phi_{NN}(\mathbf{R}) \quad = \quad \frac{1}{2} \sum_{ij} \frac{Z_i Z_j}{|\mathbf{R}_i - \mathbf{R}_j|} \tag{5.6}$$

is assumed to be included in all the energy expressions. If the system is periodic, a sum over all periodic images is introduced and sum evaluated using Ewald method [45].

Lastly, for simplicity, we have written the electron-nuclear interaction as a local function $v^{ext}(\mathbf{r}; \mathbf{R})$ only. In practice, it is a non-local term beyond the scope of the current discussion, but is discussed later when parallelization is described.

5.2.2   DFT Computations within Basis Sets

In the evaluation of DFT, it is useful to express the KS electronic states in terms of a set of known, closed form mathematical functions called a basis set

$$\Psi_s(\mathbf{r}) \;=\; \sum_k c_{sk}\chi(\mathbf{r}) \tag{5.7}$$

with c_{sk} as the expansion coefficients. In this section, we will concentrate on the application of the plane-wave basis set,

$$\Psi_s(\mathbf{r}) \;=\; V^{-1/2}\sum_{\mathbf{g}} \tilde{\Psi}_s(\mathbf{g})\exp\left(i\mathbf{g}\cdot\mathbf{r}\right) \tag{5.8}$$

with $\tilde{\Psi}_s(\mathbf{g})$ as the plane-wave basis set coefficients.

The advantages of the plane-wave basis set include: i) it is a complete, orthonormal set ensuring smooth convergence, ii) the plane-wave basis functions do not depend on atom center position which obviates basis set superposition error of Gaussian approaches and iii) in numerical simulations, fast Fourier transforms (FFTs) can be used to evaluate many of the terms, greatly increasing computational efficiency. Its main disadvantage is that it scales like N^3 and it is not easy to develop effective $O(N)$ scaling approaches with plane-waves.

5.2.3   Molecular Dynamics

In the molecular dynamics method (MD), Newton's equations of Motion in Hamiltonian form for a set of N atoms (or nuclei) [70, 3]

$$H(\mathbf{P},\mathbf{R}) \;=\; \sum_i \frac{P_i}{2m_i} + \phi(\mathbf{R}) \tag{5.9}$$

$$\dot{\mathbf{R}}_i \;=\; \frac{\mathbf{P}_i}{m_i}$$

$$\dot{\mathbf{P}}_i \;=\; F_i = -\nabla_i\phi(\mathbf{R})$$

are solved numerically on a computer. As the equations of motion themselves can be solved in linear scale (with the number of atoms, N), the scaling of the MD method is determined by the scaling of the force evaluation. In the field, the term molecular dynamics is reserved for cases where the atomic/nuclear forces are derived from a closed form empirical potential function, $\phi(\mathbf{R})$, which is usually assumed to model well the Born-Oppenheimer electronic surface. MD potential functions are often complex but can usually be evaluated with $\mathcal{O}(N)$ computational complexity.

5.2.4   Ab initio Molecular Dynamics and CPAIMD

One simple way to envision *ab initio* molecular dynamics within the DFT ES structure picture is to simply replace the empirical potential function with the minimized DFT functional. This approach is referred to as "Born-Oppenheimer" Molecular Dynamics (BOMD) [41]. That is, one freezes the atoms, minimizes the desired density functional to an appropriate tolerance, evolves the atoms one time step forward with the nuclear forces determined from the (numerically/nearly) minimized functional $E[n^{min}(\mathbf{r}); \mathbf{R}]$ and repeats. The BOMD approach is widely employed but the minimization tolerance must be taken small or Hamilton's equations can become unstable. More sophisticated versions of this procedure that fold the minimization procedure into the numerical integration so as to preserve symmetry properties of Hamilton's equations [190] are beyond the scope of this chapter.

The approach we shall adopt here is the extended Lagrangian method pioneered by Car and Parrinello [32]. The coefficients of the basis set expansion coefficients of the KS orbitals are introduced as dynamical variables along with the nuclear degrees of freedom

$$\mathcal{L}_{CP} = \frac{\mu_{faux}}{2} \sum_{s\mathbf{g}} \dot{\tilde{\Psi}}_s^2(\mathbf{g}) + \frac{1}{2} \sum_i m_i \dot{\mathbf{R}}^2 - E[n(\mathbf{r}); \mathbf{R}] - \phi_{NN}(\mathbf{R})$$

subject to the constraints

$$\sum_{\mathbf{g}} \tilde{\Psi}'_s(\mathbf{g}) \tilde{\Psi}_s(\mathbf{g}) = 2\delta_{ss'} = O_{ss'} \tag{5.10}$$

A set of the Lagrange multipliers that preserve the (holonomic) orthonormality constraint of the KS states are introduced. Using the extended "Car-Parrinello" Lagrangian, equations of motion for the simultaneous evolution of the nuclei and the basis set coefficients can be derived.

$$\frac{d}{dt} \frac{\partial \mathcal{L}_{CP}}{\partial \dot{\tilde{\Psi}}_s} - \frac{\partial \mathcal{L}_{CP}}{\partial \tilde{\Psi}_s} = 0 \tag{5.11}$$

$$\frac{d}{dt} \frac{\partial \mathcal{L}_{CP}}{\partial \dot{\mathbf{R}}_i} - \frac{\partial \mathcal{L}_{CP}}{\partial \mathbf{R}} = 0$$

If the basis set coefficients are assigned a "faux mass" parameter μ_{faux} that is sufficiently small, the initial faux kinetic energy in the basis set "coefficient velocities" is taken sufficiently small and the density functional is initially minimized, then an adiabatic separation can be invoked such that the basis set coefficients will evolve dynamically so as to keep the functional nearly minimized as the nuclei slowly evolve. Well understood MD techniques called Shake and Rattle [70] can be used to enforce the orthogonality constraints on the basis set coefficients.

5.2.5 Path Integrals

While MD and/or CPAIMD yield the motion of classical nuclei on the Born-Oppenheimer surface, this is sometimes insufficient to generate an adequate picture of the physics of a given system of interest. We shall consider two improvements in this chapter—path integral methods to add nuclear quantum effects on the ground BO electronic surface at the level of Boltzmann statistics and Parallel Tempering (Replica Exchange) to increase statistical sampling in systems with large energy barriers separating stable thermodynamic states (e.g., rough energy landscapes).

Feynman's path integral picture of quantum statistical mechanics [63] in the Boltzmann limit is particularly well suited for combination with CPAIMD. In Feynman's method, the single atom of classical mechanics is replaced by a classical ring polymer of length P beads held together with harmonic nearest-neighbor links. The classical limit arises when $P = 1$ and as P approaches infinity the results converge to the true quantum limit; the basic path integral method converges as P^{-2} and for most systems of interest $P \leq 64$ will suffice. Under Boltzmann statistics, each bead in the chain is assigned a number and only beads of different atoms with the same index interact with the 1 Pth of the potential $\phi(\mathbf{R}_i)/P$ with i the bead index. This picture is referred to as the classical isomorphism. Using advanced MD methods, it is possible to perform accurate PIMD simulations [238].

The CPAIMD method is easily grafted upon the PIMD technique to create CPAIPIMD. Simply put, the path integral method requires P electronic computations to generate $E[n_i(\mathbf{r}); \mathbf{R}_i]$ from which nuclear forces can be derived (e.g., the functional replaces the empirical potential of PIMD in a similar way as the same replacement takes MD to CPAIMD) and the CP Lagrangian can easily be extended to accomplish this change. The CPAIPIMD method can be parallelized effectively as the electron structure computations do not interact directly; quantum effect arise indirectly from the harmonic forces confining the ring polymer of each atom into a small blob ("wave-packet"). The more "quantum" the particle, the wider the spread of the beads which for example allows the isomorphism to treat quantum tunneling. Parallelization is discussed in more detail in later sections.

5.2.6   Parallel Tempering

In systems with rough energy landscapes, barrier crossing events become sufficiently rare that the results of a simulation study may not reflect the underlying physics. A system may simply become kinetically trapped in a local (free) energy minima and hence not "visit" the important regions of phase space. The same physical system at elevated temperature may, however, "traverse" phase space quite readily. In parallel tempering MD simulations (PTMD), M identical independent physical systems are run simultaneously at a set of temperatures T_i. Every fixed number of MD steps, nearest neighbors in temperature space attempt to swap temperature with probability

$$P = \text{Min}\left[1, \exp\left(\delta\beta_{ij}\delta H_{ij}\right)\right] \qquad (5.12)$$

This can be shown to lead to M properly sampled systems at the M specified temperatures [55]. There are some formal difficulties using constant temperature MD methods to drive the dynamics of the M systems but these are considered minor and PT-MD is a well established method. It is most simple to use BOMD to implement PT within *ab initio* techniques and this is the course we are currently pursuing. For parallel computations, PT-BOMD is quite attractive as the M BOMD simulations rarely communicate, and when they do, they need only exchange energies and temperatures.

5.3   Parallel Application Design

OPENATOM was envisioned from its inception as a fine-grained implementation of Car-Parrinello *ab initio* MD using Charm++ as its parallelization substrate. Prior work, by our collaborators Glenn Martyna and Mark Tuckerman, in developing the PINYMD physics engine had already overcome the challenges of method implementation and validation. Hence, we elected to integrate the sequential simulation components from PINYMD into the design of OPENATOM. This resulted in a two level design with parallel control structures, Parallel Driver, implemented in Charm++ making calls to the integrated PINYMD routines implemented in C++ and Fortran.

5.3.1   Modular Design and Benefits

The overall CPAIMD algorithm is composed of several data manipulation and computation steps. One can envision the electronic states and the eventual atomic forces as the fundamental data that are computed and evolved through the simulation. Typical numerical algorithms express computations as a sequence of steps that operate on input data. Parallelization occurs by simply dividing large volumes of this data into smaller pieces. This naturally yields a procedural, bulk-synchronous expression of the algorithm suitable for coarse-grained weak scaling.

However, OPENATOM achieves its fine-grained parallelization by identifying the major steps of the algorithm and expressing each piece of the computation separately. Chare classes encapsulate the logic needed for each such piece of the computation along with the state needed for that piece across multiple iterations. The data of interest (electronic states, forces, etc.) then simply flow back and forth across these pieces as they evolve over the simulation. The different pieces of the algorithm are wired together by directing the output of a piece (class) as a message that triggers the computation in the next piece. This takes the form of remote method invocations. Traditional data decomposition is also trivially expressed by partitioning the input, output and computations in each piece of the algorithm (chare class) across many instances (objects) of that piece. There are several benefits that stem from such a design exercise.

- Parallelism arises from both a functional decomposition of the computation and the traditional decomposition of the data being operated upon.

- Computations are driven by the data sender. Data receivers do not have to post receives, or take any preparatory actions to receive this data. In fact, objects that receive data (or messages) do not have to be aware of type, location or even the existence of a sender. This semantic promotes looser coupling of interacting pieces in the software. Class interfaces separate components that produce or consume each other's data. Such approaches are not new, but well-trodden paths in other domains. The challenge in HPC has been to convince the larger community that high performance can tolerate such loose coupling and other software engineering ideals. Frameworks like Charm++ help scientific applications realize performance without sacrificing modular and maintainable software.

- Loose coupling also permits easy selection of different numerical methods/functionality by simply instantiating the objects of the appropriate chare classes. As long as they provide the same interfaces and data guarantees, the remaining application is unmodified.

- The messaging model in Charm++ permits modifications to the communication structure of an individual parallel component without concerns for introducing subtle parallel bugs like deadlocks or races. Charm++ does not preclude such issues completely, but only mitigates the need to understand the global communication state (all sends, receives, etc.) before introducing other parallel communication.

- Unit testing for numerical software is somewhat challenging. Correctness or failure may be determined far to the right of the decimal point! Usually, in numerical algorithms only the initial inputs and the final output can be easily accessed or validated. However, the loose coupling described above makes it easier to test individual software components. Mock environments and test harnesses are easily set up to interact with isolated components of a large, parallel application. OPENATOM has used this capability on several occasions to detect regressions and bugs in individual parallel components in a setting that is independent of the remaining application.

- Software components, in our experience, have differing rates of change. Some pieces are very stable and only need minor, occasional tweaks. Other parts experience constant modifications, enhancement or tuning. We have found this true of OPENATOM too. We found it helpful to introduce parallel interfaces at these layers of shear between differing rates of evolution. By isolating rapidly changing parallel components behind chare interfaces, we were able to insulate domain logic and other parallel modules from refactoring.

5.3.2 Parallel Driver

OPENATOM's parallel driver is composed of classes that match the logical expression of the terms of the Car-Parrinello method, supplemented by classes for optional numerical methods and features. Many of these classes that represent key steps of the computation or key representations of the data are promoted to chares classes with entry methods that clearly represent phases or stages of the computation. Instances of these classes are usually collected into chare arrays across one or more dimensions.

5.3.2.1 Decomposition

Multi-dimensional chare arrays are the primary expression of decomposition. The plane-wave pseudo-potential is expressed in terms of electronic states; each state is a 3D collection (usually a rectilinear box) of points. To facilitate a decomposition finer than the number of electronic states, a slice along one dimension is performed to form planes, which produces a natural decomposition along the dimensions of states and planes.

Let us preface our discussion with the following terms:

S : the number of electron states

N_c : the number of chunks of gspace plane-wave

N_d : the number of chunks of gspace density

P : the number of planes in the x-dimension of real space plane-wave

Sub : subplanes of decomposition for electron density

A : number of atoms

n_{atype} : number of atom types

The primary chare classes involved in expressing the CPAIMD algorithm in OPENATOM are listed below:

GSpace Driver (2D chare array $[N_c \times S]$) Handles flow of control within an instance, always same dimensional cardinality and mapping as gspace.

Electronic State Plane-wave GSpace (2D chare array $[N_c \times S]$) Handles the electronic structure in Fourier space (referred to as GSpace hereafter). Due to spherical cutoff sparsity, GSpace is rearranged into approximately equal size chunks. The number of chunks N_c is a free parameter selected at runtime.

Electronic State Plane-wave Real Space (2D chare array $[P \times S]$) Handles electronic structure in real space. The points of plane-wave pseudo-potential are cut along the x-dimension for finer parallelization.

Electron Density Real Space (2D chare array $[P \times Sub]$) Handles electron density in real space. Each plane may be further subdivided into subplanes at runtime for additional parallelism.

Electron Density GSpace (1D chare array $[N_d]$) Handles electron density in Fourier space. Due to spherical cutoff sparsity, GSpace is rearranged into approximately equal size chunks. The number of chunks N_d is a free parameter selected at runtime.

Electron Density Real Space Hart (3D chare array $[P \times Sub \times n_{atype}]$) Handles electron density hartree computation in real space.

Electron Density GSpace Hart (2D chare array $[N_d \times n_{atype}]$) Handles electron density hartree computation in Fourier Space.

Atoms (1D chare array $[A]$) Handles atomic positions, velocities and corresponding data for computation of forces and positions.

AtomsCache (chare group) Provides globally available cache of positions and forces.

Non-local Particle GSpace (2D chare array $[N_c \times S]$) Handles non-local particle force computation Fourier space. Always same dimensional cardinality and mapping as GSpace.

Non-local Particle Real Space (2D chare array $[N_{nlees} \times S]$) Handles non-local particle force computation real space. N_{nlees} is determined by the x-dimension of the EES grid.

Orthonormalization (2D chare array $[S_{nog} \times S_{nog}]$) Handles orthonormalization based on iterative inverse sqrt method. S_{og} is set at runtime to be a factor of S_{pg}, $O_{nog} = S/O_{og}$.

Electron Pair Calculator (4D chare array $[P \times S_{npg} \times S_{npg} \times C]$) Computes the electron state pair matrix multiplication and correction for electronic structure plane-wave forces and coefficients. S_{pg} is set at runtime to be a fraction of S. $S_{npg} = S/S_{pg}$.

Euler Exponential Spline Cache (chare group) Provides globally available cache for EES.

Structure Factor Cache (chare group) Provides globally available cache for structure factor.

5.3.2.2   Control Flow

Each major category of objects (see Figure 5.1): GSpace, RealSpace, Density, etc. has a distinct flow of control. That flow is expressed in the RTH Thread suspend/resume syntax extension of the Charm++ RTS. The flow is implicitly expressed by progress in program order through an event loop, wherein dependencies are explicitly expressed by application condition vari-

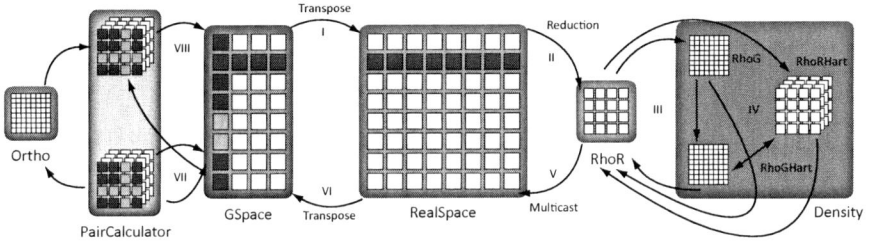

FIGURE 5.1: The important chare arrays in OPENATOM that provide both data and functional decomposition. Major interactions between the chare arrays are depicted with arrows. The phases of computation and the typical order in which phases are triggered within a step of the simulation are indicated by the numbering of these interactions. Many of these interactions embody collective data movement operations from or to slices of chare arrays. Some examples of sections (slices) of chares that communicate with each other are indicated by the identically colored objects in the RealSpace, GSpace and PairCalculator chare arrays.

able tests that guard local method calls on the chare, along with remote method invocations to communicate results to other objects. Object methods implement the computation for each phase of the computation and resume to the event loop, or the Charm++ RTS scheduler, as necessary. Entry point methods on the chare will set the relevant condition variables for the arriving data and resume into the event loop.

The key advantage of this scheme is that it provides a clear encapsulation of the state of each object, as the intersection of program order and tests of condition variables enforce algorithmic constraints for each object. This synergizes with Charm++'s support for adaptive overlap by allowing each object to safely progress through its state independently, as its constraints are met. It also allows for further application tuning, whereby computation can take place as early or late as the critical path of the application dictates. Furthermore, the communication of results from those computations can be throttled, or expedited, as appropriate.

For example, the FFT operation from *GSpace* to *RealSpace* results in a many to many personalized communication pattern, where each *GSpace* object issues a force message to each *RealSpace* object which shares its electronic state, that contains the portion of the result of the local FFT transform corresponding to the destination's plane index. *GSpace* objects then suspend until the Ψ x VKS data is returned from *RealSpace*. Meanwhile, *RealSpace*'s event loop collects FFT inputs until all have arrived, then completes the transform and initiates a reduction to sum across the states to produce one result for each plane of *Electron Density* objects.

Simultaneously with the previous paragraph's activities, the *Non-local* computations are overlapped with the *Electron Density*, with the latter taking priority. Due to the fact that number of data elements in the electron state grid is typically much larger (by at least an order of magnitude) than the electron density grid, the former has significantly better strong scaling characteristics than the latter. The automatic prioritized overlap allows computation units to do either, or both, at various strong scaling decompositions. This allows the application to efficiently scale up to a larger number of computational units before the serial bottleneck from the *Electron Density* dominates performance.

As shown in Figure 5.1 (electron state and density phases are shown) the control flow is data dependency directed to evolve the electron states, reduce them for the electron density, initiate the non-local force computation and integrate nuclear forces based in each step. Decomposing the problem into distinct chares for each phase of the computation allows the implementation and placement decision for each element to be taken independently, or to build upon choices made for related phases, as necessary. See Figure 5.2 for how these decomposition options are applied in the context of network topology aware mapping.

5.3.2.3   Multiple Interacting Instances

The above comprise the primary components required to simulate one instance of a molecular system. In addition, several simulation capabilities require multiple interacting copies (instances) of a molecular system within the same simulation. These are handled by instantiating multiple copies of the above chare collections. Each instance hosts a set of the above interacting components wired to receive and send computation output to each other. All such components within an instance share a common identifier called the UberIndex.

The UberIndex is a higher level of organization imposed upon the instances of the chare arrays to implement features requiring the interaction of multiple variants of a system, such as Path Integrals, K-Points, Tempering and Spin Orbitals. Each UberIndex contains an instance of each of the chare array classes and forms a complete description of a target molecular system. Coordination across UberIndices occurs in accordance with the synchronization required by each feature and users may use between zero and all Uber features, such as both Path Integrals and K-Points, in the same simulation.

Let these terms quantify UberInstance selection on decomposition:

I : number of instances

T : number of temperatures

B : number of path integral beads

Kp : number of k-points

S : number of spin directions (when enabled this is 2 for up and down).

Simulations based on multiple instances require the following additional chare classes:

Instance Controller (1D chare array [I]) Handles the instance initialization and coordination of cross instance communication. Handles coordination specific to cross B, cross Kp, cross S and across all I.

Path Integral Bead Atom Integrator (1D chare array [A]) Handles integration of atom positions across path integral beads.

Temperature Controller (1D chare array [T]) Handles exchanging temperatures across Tempers.

5.3.3   Topology Aware Mapping

The Charm++ runtime maps various chare arrays in OPENATOM to the physical nodes and cores automatically. This default mapping is load balanced but possibly not optimized with respect to the specific communication patterns in OPENATOM. The runtime gives the freedom to the application developer to decide the placement of the chare arrays. Since OPENATOM is communication-bound, a load balanced mapping aimed at minimizing the inter-node communication was developed. Even with this optimized mapping, OPENATOM suffered from performance problems. Performance analysis on a large number of processors hinted at network contention problems due to heavy communication. To mitigate network contention, we started exploring interconnect topology aware mappings of the communication-bound phases in OPENATOM. Topology aware mapping aims at reducing the number of hops/links traversed by messages on the network to minimize link sharing and hence contention. This is achieved by placing objects that communicate frequently close together on the physical network.

Figure 5.1 presents the important phases and chare arrays in OPENATOM and the communication between them. The two-dimensional (2D) GSpace array communicates with the 2D RealSpace array plane-wise through transpose operations. The same GSpace array communicates with the 3D PairCalculator array state-wise through reductions. Optimizing one communication requires putting all planes of each state in GSpace together whereas the other communication benefits from placing all states of each plane in GSpace together. A hybrid approach that balances and attempts to favor both communications has been developed. There are other communications between RealSpace and the density chares and ortho and the PairCalculator chares which also need to be considered.

Heuristics that optimize both the GSpace $\longleftrightarrow$ PairCalculator and GSpace $\longleftrightarrow$ RealSpace communication were considered and mappings of these chare arrays to three-dimensional torus networks were developed [21]. Figure 5.2 shows the mapping of these three chare arrays to a 3D torus partition. The

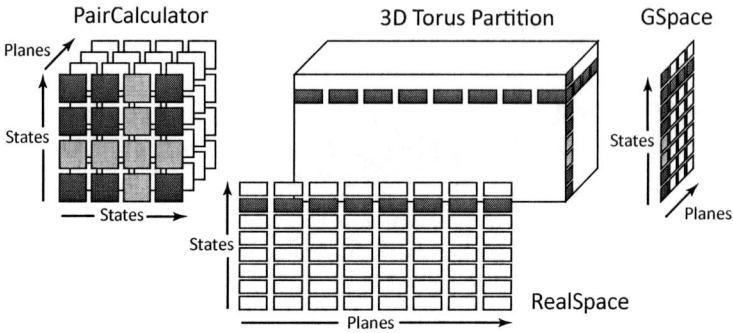

FIGURE 5.2: Mapping of GSpace, RealSpace and PairCalculator chare arrays to the 3D torus of a Blue Gene machine.

GSpace array is mapped first to prisms (sub-tori) in the allocated 3D job partition. The RealSpace and PairCalculator chares are then mapped proximally to the GSpace array. This leads to significant performance improvements as shown in Figure 5.3.

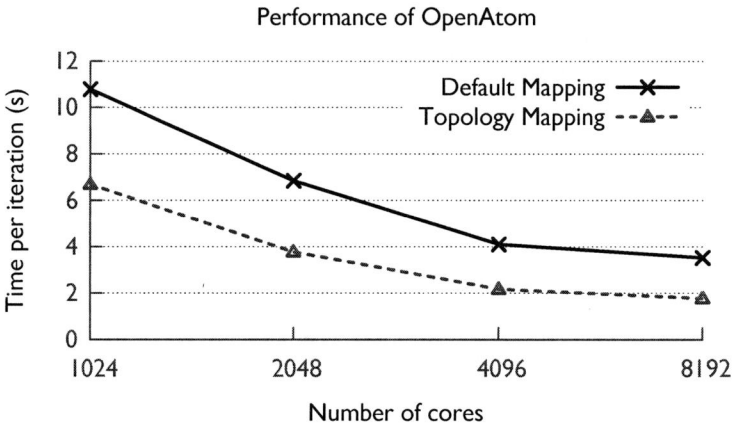

FIGURE 5.3: Performance improvement of nearly two times on 8192 cores of Blue Gene/P by using a topology aware mapping of OPENATOM (WATER_256M_70Ry).

5.4   Charm++ Feature Development

OPENATOM has driven the development of Charm++ in several ways. It is a case study in unified data and functional decomposition, and has also driven many capabilities required to support chare collections that span only a subset of the total number of processors in an execution. Some of the features in Charm++ that were inspired or partially driven by the requirements of OPENATOM are listed below:

Static Balance OPENATOM has no inherent dynamic load imbalance. It achieves its performance benefit from Charm++ due to fine-grained decomposition and the automatic overlap of prioritized computation phases. This demonstrates the unadulterated benefits of these techniques, in contrast to other Charm++ applications which use dynamic load balancing. The aggressive use of prioritized messages in OPENATOM has been a driving use case for the development of a robust and efficient runtime implementation. Additionally, Charm++ also sports runtime and build time options that can turn off dynamic load balancing and other instrumentation required by dynamically evolving applications.

TopoManager Library The plane-wise communication phases alternating with state-wise communication phases, along with the independent expression of these operations in distinct chare arrays, demonstrate a high sensitivity to topology aware placement. This drove the development of the robust, flexible, cross platform TopoManager library that exposed the underlying network topology of the system across supercomputers with torus networks from different vendors.

CkMulticast Library The communication between Electronic State Planewave GSpace and the Electron Pair Calculator is confined to each plane. When executing at scale, each chare will typically have tens of data exchange partners, each of which must receive a part of its electronic state, and the entire state must be updated, returned and reassembled. This has driven the development of the CkMulticast library to provide an efficient infrastructure to support the multicast and reduction operations with pipelining, customized control of spanning tree width, prioritization and fragmentation. Specifically, OPENATOM was a heavy user of operations involving sections of a chare array (a plane of GSpace chares or a prism of Pair Calculator chares). This drove the optimization of multicasts and reduction to chare array sections.

Topology-aware multicasts and reductions OPENATOM's decomposition and design requires many common data movement patterns (multicasts, reductions, scatters, all-to-all). Several of these are expressed as operations involving a regular slice (section) of a chare array. For example data might need to be multicast from a plane of GSpace objects to a prism of Pair Calculator objects. However, the actual communication required does not translate cleanly to the underlying processors. This is because the number of objects is influenced by the problem and grain sizes, and their placement is influenced by topology and load balance considerations. OPENATOM, hence, performs many multicast and reduction operations that typically translate to a clustered but arbitrarily shaped subset of processors within the overall network topology. This is also, and especially, true for other Charm++ applications that require object migrations to effect load balance.

In the strong scaling regime of the execution spectrum, OPENATOM is fairly sensitive to communication and messaging behavior. The fine-grained parallelization results in a greater emphasis on optimized data movement. Thus OPENATOM created a use case and drove the implementation of network-topology awareness in the Charm++ implementations of the multicast and reduction operations. This is implemented in Charm++ via the construction of topology aware spanning trees. A detailed description of this implementation is beyond the scope of this text.

Figure 5.4 illustrates the speedup obtained by OPENATOM when using a version of the Charm++ runtime system that could dynamically construct topology aware spanning trees over subsets of processors. The speedups are relative to the performance of OPENATOM at each of those processor counts without the use of topology aware spanning trees. We note that the application performance improves considerably. All Charm++ applications that perform multicasts and reductions over chare array sections now benefit from topology-aware multicasts and reductions whenever topology information is available.

Arrays Spanning a Subset of the Processor Allocation The most computationally heavy phases in the simulation scale to 20x the number of elec-

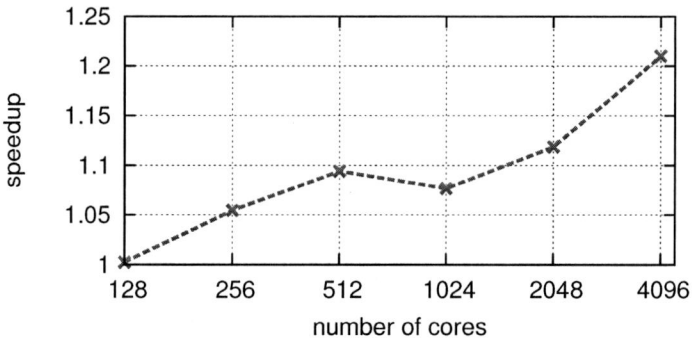

FIGURE 5.4: Speedup of OPENATOM due to topology-aware multicast and reduction operations in Charm++ relative to base cases without such topology aware collective communication. Data was obtained on Surveyor, an IBM Blue Gene/P, for the WATER_32M_70Ry dataset.

tronic states. However, several phases on the critical path, such as orthonor-malization and electron density calculations, have portions which cannot prof-itably be decomposed as finely as the rest of the algorithm. The chare arrays for these phases will have fewer elements than the number of processors. OPE-NATOM has driven, and continues to drive, optimizations in the Charm++ runtime system for efficient support of such arrays. This requirement has also provided the initial basis for the later development of the UberIndex scheme, wherein all arrays span exclusive subsets of the processor allocation.

Many Multidimensional Chare Arrays The plethora of chare arrays and the total number of chare objects drove the development of robust and efficient support for the construction of many millions of objects on terascale machines.

5.5 Performance

OPENATOM scales well on the IBM Blue Gene architecture series, as shown in Figures 5.5 and 5.6. The high communication intensity of the algorithms benefits from the relatively balanced approach in the design of the Blue Gene series. Each figure shows the strong scaling performance of benchmark systems composed of molecules of liquid water with a 70 Rydberg cutoff at the Γ point, ranging from 8 to 256 molecules on the Blue Gene/L, and confined to 32 and 256 molecules on Blue Gene/P and Blue Gene/Q. All the plots use log_2 on the X-axis and log_{10} on the Y-axis.

Figure 5.5 presents the time per iteration for six different water systems ranging from 8 to 256 molecules. All the runs were done in the co-processor

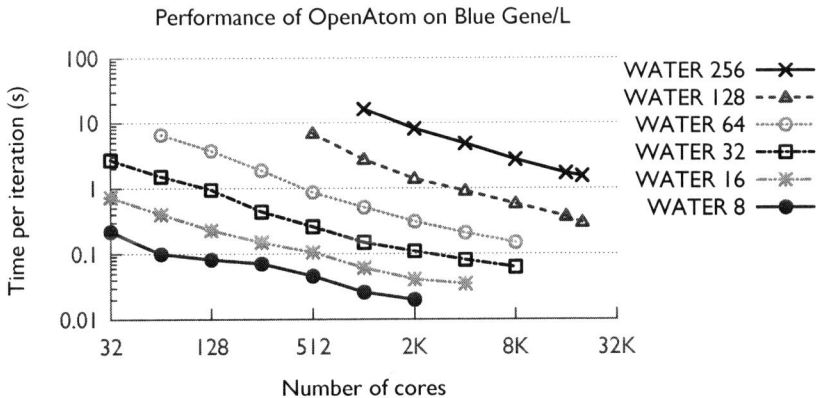

FIGURE 5.5: OPENATOM on Blue Gene/L (CO mode).

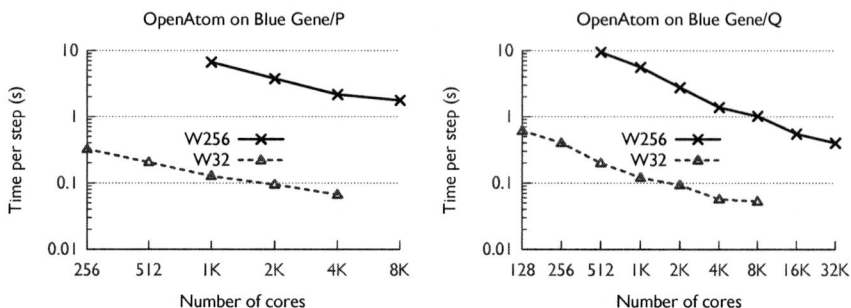

FIGURE 5.6: OPENATOM on Blue Gene/P (VN mode) and Blue Gene/Q (c16 mode).

(CO) mode which only uses one processor on each Blue Gene/L node for computation. To consider weak scaling performance, it must be noted that due to the dominance of $O(N^3)$ methods with increasing system size, doubling the number of molecules corresponds to an eight-fold increase in the amount of work. Therefore the corresponding order of magnitude relative time per step performance of these benchmarks represents good weak scaling. The largest 256 water molecules system scales well to 32,768 processors.

The left and right plots in Figure 5.6 show the scaling performance on Blue Gene/P and Blue Gene/Q, respectively. The runs on Blue Gene/P were done in virtual node (VN) mode, i.e., placing one process on each physical core. The runs on Blue Gene/Q were done in a similar on Blue Gene/Q referred to as the c16 mode. Once again, we see good scaling behavior for the larger benchmark system up to 32,768 cores of Blue Gene/Q.

5.6   Impact on Science and Technology

The OPENATOMm software suite [25] has been employed to gain insight into important systems spanning chemistry, biology, physics, materials science and engineering. Here, we discuss two application studies which illustrate the ability of the CPAIMD method to generate important insights into systems of high interest in science and technology (S&T).

5.6.1   Carbon Based Materials for Photovoltaic Applications

One component of the green energy revolution involves the widespread adoption of photovoltaic (PV) cells across an array of energy applications. However, breaking into the highly competitive energy market is quite difficult and a wholly economically driven adoption of solar cell technology requires breakthroughs that will lower cell cost, increase reliability and decrease the cost of installation. This, of course, requires, in turn, innovative exploratory research spanning all aspects of S&T.

There are several strategies that could be used to power the revolution [128, 78]. One involves the use of high quality crystalline silicon (c-Si) solar cells in conjunction with heat scavenging and solar concentrators; these high efficiency systems require harvesting every last bit of energy to offset the high (fixed) cost of c-Si. On the low end, organic PV cells can be cheaply ink-jet printed, need large areas due to low efficiency and are difficult to fabricate with the 25-30yr lifetimes required in some applications. An intermediate strategy is to build amorphous silicon (a-Si) solar cells which are stable to long times, and are sufficiently cheap to avoid concentrators and scavenging strategies, and would be adopted widely if costs could be dropped.

In Figure 5.7, a mock-up of an a-Si solar cell is given. Unlike c-Si solar cells, a top transparent conducting material or transparent conducting electrode (TCE) is required to conduct electrons away as doped a-Si mobilities are too low for the cell to function well. In current a-Si solar cell designs, the TCE is made of an Indium-Tin-Oxide (ITO) alloy. While these oxide materials perform very well, they are expensive both in terms of both materials (Indium) and processing (high temperature deposition of a metal oxide). It would be therefore an important advance if the ITO could be replaced by a cheaper, more easily processed material.

It has been suggested that graphene, a zero band gap insulator, which is a single carbon atom thick, would, when doped, make a very effective TCE (see Figure 5.8); that is, graphene is both thin and hence transparent and yet highly conductive when doped [131]. The recent development of a copper based process that yields high quality, large area graphene sheets [156] makes

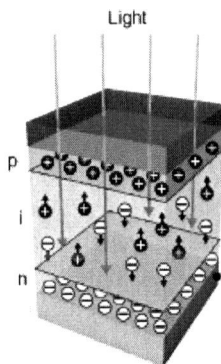

FIGURE 5.7: Mock-up of a solar cell with a transparent top electrode, a PIN junction and a bottom electrode (**see Color Plate 5**).

FIGURE 5.8: Graphene is a single atomic layer thick material made entirely of carbon atoms. Graphite, commonly used in pencils, is formed by stacking many layers of graphene.

investigation of the properties of graphene in the context of solar cell TCE's very pertinent.

We have employed the CPAIMD technique as embodied in the OPENATOM software package to examine the physics and chemistry underlying the doping of graphene sheets. The basic physics we wish to observe is called rigid band doping. That is, the band structure (and hence density of states) does not alter in the low energy regime (near the Fermi level); the dopant merely serves to inject electrons/holes and hence shift the Fermi-level. Rigid band doping increases the number conduction pathways through the materials and hence the conductivity without diminution of charge carrier mobility through scattering; the intrinsic mobility of carriers in graphene is quite high.

In our studies of graphene doping, we considered a non-volatile dopant, $SbCl_5$, which would be expected to have long lifetime in a solar cell (Figure 5.9). We observed that simply setting up a regular lattice of $SbCl_5$ molecules on the graphene surface did not cause a shift in the Fermi-level (e.g., no rigid band doping observed). That is, simply setting up a (guessed or presumed) structure and performing energy minimization did not properly predict the experimentally observed physics. We next allowed the molecules to evolve naturally according to Newton's equations using CPAIMD. We observed that the $SbCl_5$ molecules spontaneously dissociated following the disproportionation chemical reaction,

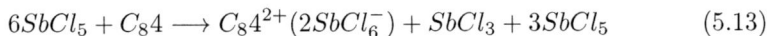

$$6SbCl_5 + C_84 \longrightarrow C_84^{2+}(2SbCl_6^-) + SbCl_3 + 3SbCl_5 \qquad (5.13)$$

The open shell products p-dope induce rigid band doping of the sheet [189] (the system studied contains 6 molecular entities placed between two graphene layers containing 84 carbon atoms each).

The observed $SbCl_5$ disproportionation chemistry is catalyzed by

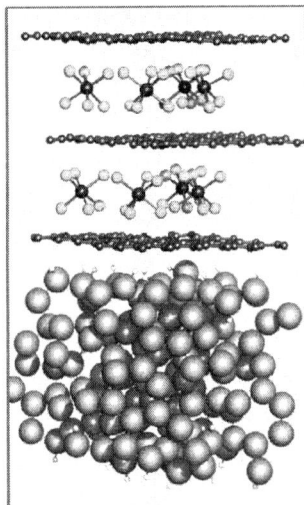

FIGURE 5.9: A snapshot of a simulation of (from bottom to top) amorphous silicon terminated by hydrogen atoms, graphene, a layer of antimony pentachloride (SbC 15), graphene, a layer of antimony pentachloride (SbCl$_5$), graphene **(see Color Plate 5)**.

graphene, itself, which functions as a metal-surface. There are antecedents in the graphite intercalation compound literature (e.g., graphite is essentially an infinite number of A-B stacked graphene layers). However, in order to form intercalation compounds, extremely harsh conditions are applied to allow the compounds to enter/intercalate into the graphite lattice, making it unclear how the disproportionation reaction takes place. Our computations show clearly the mechanism involves a (metal) surface catalyzed charge transfer reaction that occurs spontaneously.

In order to further decrease the sheet resistance of the graphene, we have designed and tested a screen-printed busbar pattern experimentally. In order to determine the best metal for the design, we have used *ab initio* methods to study the physics and chemistry that leads to a high performance metal-graphene interface (e.g., emits a high tunneling current). Our current work involves studies of the a-Si-graphene interface which is key in developing an integrated solar cell solution.

5.6.2   Metal Insulator Transitions for Novel Devices

Although Moore's law continues in computer technology (chip features shrinking exponentially quickly with time), Dennard scaling, which allows clock frequency to increase concomitantly with feature size decrease, ceased abruptly in 2003 [236]. This halt is not due to a failure of engineering processes but occurs because CMOS, the current chip technology, has reached hard limits imposed by physics. Basically, preventing a charge carrier from crossing an electrostatically gated barrier requires in practice at least $\approx 1\mathrm{V}$ line voltage at the operating temperatures and length scales of current technology. The computer industry therefore is in need of new approaches to digital switching that employ different physics.

Previous exploratory device research at IBM involved using electron-injection gated metal-insulator transitions (MIT) to provide the required switching (1=conducting, 0=insulating) physics [188]. The idea is to poise the channel material near the MIT, and provide just enough charge injection gating to push the system from the insulating to the conducting regime. This concept is, in fact, quite general and powerful, and not limited to charge injection mechanisms for the MIT.

We have also performed exploratory scientific research at IBM on pressure driven metal insulator transitions to study switching mechanisms of Phase Change Materials, specifically germanium doped Antimony, Ge_xSb_{1-x}, $x = 0.15$ [217]. In a typical application involving phase change materials, a heat pulse is applied to the conductive crystalline phase which creates an amorphous insulator and a more moderate annealing pulse is applied to the amorphous form to switch the material back to its crystalline state. This physics forms the basis of a non-volatile memory technology called Phase Change Memory. We have compared and contrasted pressure switching of the GeSb material [217]. Simulations of heating annealing and pressure annealing of the material yield similar end products. We concluded that the mechanism for the transition was a phenomenon termed gap-driven amorphousization. As the crystalline material is heated or put under tensile stress, it begins to become favorable for the electronic band gap to open so as to lower free energy (as temperature or tensile stress has increased). As the gap opens, the bonding pattern changes (more 4-coordinate defects appear), the crystalline order decreases and the material evolves into an amorphous state. The amorphous state can be placed back into the crystal by applying compressive stress. The calculations were performed on 192 atom systems (29 Ge and 163 Sb) for very long times (100 picoseconds per quench).

We have recently explored combining the two approaches to create digital switches using pressure driven metal insulator transitions [186]. In order to make a switch as opposed to a memory, materials such as the intermediate valence compound SmSe that undergo a *continuous* MIT, a decrease of 4 orders in resistivity with the application of 1-2 GPa of pressure, are used. The gating is accomplished through the application of a voltage across a piezoelectric material. We have modeled our novel device which we term the Piezotronic Transistor (PET) theoretically and shown that it can be switched at very low line voltage 0.1V and yet maintain high speed (10 GhZ). If the PET can be successfully fabricated at the nanoscale, it would represent an important new technology. We are currently pursuing the experimental embodiment of this device [186, 187].

5.7   Future Work

The introduction of replica style computations (Path Integrals, k-points, parallel tempering and combinations thereof) has greatly increased the scale of machine, and the kind of simulation experiments, that the OPENATOM software can support. Path integrals allow computations of systems where nuclear quantum effects are important such as hydrogen exchange reactions for instance. Replica exchange permits increased sampling in systems with rough energy landscapes such as corrugated surfaces and biomolecular config- urations. Including k-points allows increased accuracy even for large systems including metals which are important for the study of novel electronic devices. We are just beginning to explore these exciting new applications at present enabled by OPENATOM.

The implementations of several highly desirable simulation features are in the planning stages. Broadening the applicability of OPENATOM will be served by adding support for GW-BSE, hybrid density functionals, fast super-soft pseudo-potential techniques, localized basis sets and CPAIMD-MM. GW-BSE will permit the study of excited properties of materials and bio-materials. The study of insulators is made more accurate by using hybrid DFT methods while super-soft pseudo-potential techniques allow systems containing transition and post-transition metals to be studied such as metal-enzyme reaction centers and rare earth chalchogenide semiconductors. Localized basis sets will allow linear scale methods to be implemented increasing the system sizes that can be studied whilst CPAIMD-MM will permit a region treated with DFT methods to be embedded in a large bath of atoms treated more simply (empirical potential functions).

Lastly, the software infrastructure underlying OPENATOM will be up- graded. The expression of flow control will be improved by refactoring to use the higher level Charisma language, which has matured towards production over the course of this project. Refactoring the current plane-wise FFT decom- position into pencil form will improve strong scaling performance. A number of improvements in Charm++ infrastructure, such as the TRAM streaming module, will be leveraged to further improve performance. These changes are expected to improve both the performance and usability of the application to expand its user community and extend the power of high performance computing to a wider variety of experimental challenges.

Acknowledgments

This work would not have been possible without the efforts of the many people who have contributed to the OPENATOM software over the years, including: Anshu Arya, Abhinav Bhatele, Eric Bohm, Chris Harrison, Sameer Kumar, Marcelo Kuroda, Glenn Martyna, Justin Meyer, Razvan Nistor, Esteban Pauli, Yan Shi, Edgar Solomonik, Mark Tuckerman, Ramkumar Vadali, Ramprasad Venkataraman, Da Wei, and Dawn Yarne.

This work began under the National Science Foundation grant ITR 0121357 and was continued under the Department of Energy grant DE-FG05-08OR23332. This work was performed under the auspices of the U.S. Department of Energy by Lawrence Livermore National Laboratory under Contract DE-AC52-07NA27344 (LLNL-BOOK-608553).

This research used resources of the Argonne Leadership Computing Facility at Argonne National Laboratory, which is supported by the Office of Science of the U.S. Department of Energy under contract DE-AC02-06CH11357.

Chapter 6

N-body Simulations with ChaNGa

Thomas R. Quinn

Department of Astronomy, University of Washington

Pritish Jetley and Laxmikant V. Kale

Department of Computer Science, University of Illinois at Urbana-Champaign

Filippo Gioachin

Hewlett-Packard Laboratories Singapore

6.1   Introduction

It is remarkable that cosmology is now a "precision" science. The age of the Universe is known to about a percent; the relative amounts of baryons, dark matter and dark energy are known to a few percent; the expansion rate is also known to a few percent. This is in spite of the fact that we know very little of the nature of the main constituents of the Universe: dark matter and dark energy. What we do know is that they gravitate, and gravity is by far the dominant force on astronomical scales. Hence, building models of large scale structure and making testable predictions is relatively straightforward: the matter is represented by a number of particles, and the motion of these particles is followed under their mutual gravitational forces, a technique referred to as an N-body simulation. Noted successes in cosmological model building include predicting that dark matter around disk galaxies are more spherical than the light distribution [193], and ruling out light neutrinos as a major component of the dark matter [245].

For many self-gravitating astrophysical systems, and certainly for all those dominated by dark matter, the number of interacting bodies in the system is much larger than can be realized on a computer. Hence, the "bodies" or particles in an N-body simulation code do not represent individual objects, but samples of a distribution function, $f(\mathbf{x}, \mathbf{v})$ where $f(\mathbf{x}, \mathbf{v})d^3x d^3v$ is the probability of finding a particle in the six dimensional phase space volume $d^3x d^3v$ located at $(\mathbf{x}, \mathbf{v})$. This distribution function obeys the collisionless Boltzmann equation,

$$\frac{\partial f}{\partial t} + \frac{\partial f}{\partial \mathbf{x}} \cdot \mathbf{v} - \frac{\partial f}{\partial \mathbf{v}} \cdot \frac{\partial \Phi}{\partial \mathbf{x}} = 0,$$

where $\Phi(\mathbf{x})$ is the gravitational potential generated by the mass distribution. The characteristics of this equation are simply Newton's equations of motion, and the sum of all the particle potentials is an approximation to the total potential. This approximation is improved by softening the potential around each particle [44], but it is only by increasing the total number of particles representing $f(\mathbf{x}, \mathbf{v})$ that relaxation effects can fully be mitigated.

Since the quality of a simulation improves with the number of particles, N, efficient solvers are needed, i.e., solvers that scale linearly, or at worst $N \log(N)$ with the number of particles. Furthermore, the attractive-only nature of gravity leads to large density contrasts, which require the solver to be adaptive as well. One of the first widely used adaptive techniques in astrophysical N-body solvers was the particle-particle, particle-mesh (PPPM) [57, 102] method, where long-range forces are determined using a grid, and short-range forces are computed using direct summation. Although the technique is somewhat limited in terms of adaptivity, simulations using this method significantly advanced our knowledge of cosmological structure formation [43, 182]. Furthermore, the use of fast Fourier transforms (FFT) to solve gravity on large scales makes these methods particularly well suited for simulating an infinite universe using a periodic volume.

Tree codes for gravity became popular following the publication of the Barnes and Hut paper [10], and this method was soon extended to include gas dynamics using the Smooth Particle Hydrodynamics (SPH) method [100]. As well as allowing for any level of adaptation, the tree method is not restricted to a particular geometry, which makes it particularly useful for isolated galaxies and clusters of galaxies. Tree codes have been extended to include periodic boundary conditions [99], perhaps most successfully in hybrid tree/particle-mesh methods [228] where the long-range gravity is solved using an FFT while the tree is used for short-range forces.

The irregular nature of the tree-walk is a significant challenge for parallel implementations. Nevertheless, good parallel performance has been achieved, as evidenced by the winning of a couple of Gordon Bell prizes by the HOT code [243, 244]. More recently, the scaling of a tree code on a GPU cluster also resulted in a Gordon Bell prize [96]. Tree codes have also been used successfully for a number of recent results in cosmology. In particular, the detailed phase space sub-structure of dark matter halos has been investigated in a number of state-of-the-art simulations [51, 227, 229] with as many as one billion particles per halo. While not as large in particle count as uniform volume simulations using the hybrid tree/particle-mesh technique [5], these simulations present a challenge because of their extreme dynamic range in time and space.

Because of the challenges of load balancing and strong parallel scaling needed to accomplish the simulations necessary for cosmological structure formation, we have investigated building a new simulation code, CHaNGa (CHArm N-body GrAvity), using the Charm++ programming system. The next section describes the design of this code, including the features necessary for state-of-the-art cosmological simulations. We present some accuracy tests in Section 6.3, both in terms of detailed force accuracy and cosmologically relevant end-to-end tests. Performance results, both on standard HPC machines and GPU clusters, are presented in Section 6.4, and we end with conclusions and areas of future work.

6.2   Code Design

Below we describe the main components of the gravity and SPH calculation within CHaNGa. Much of the design was based on our previous experience with PKDGRAV, the gravity code which provides the basis for the gravity/SPH code, GASOLINE [240]. In particular, the tree-walk follows an algorithm developed in [230] and the routines for calculating and evaluating multipole moments are directly from PKDGRAV. On the other hand, the domain decomposition and load balancing are unique to CHANGA since they rely strongly on the capabilities of the Charm++ runtime system.

6.2.1   Domain Decomposition and Load Balancing

Domain decomposition is a critical component of achieving parallel performance, but it requires satisfying somewhat conflicting requirements. These include a reasonable distribution of data across processors and an even distribution of computational work (load balancing). At the same time interprocessor communication needs to be minimized. These various constraints can be difficult to satisfy simultaneously, particularly for a highly non-uniform distribution of particles. For example, in using a tree structure to decompose a galaxy where the particles are centrally concentrated, an optimal K-D tree, while guaranteeing equal numbers of particles in each domain, will produce nodes with very high aspect ratios. These nodes, with at least one large dimension, fail to satisfy a multipole accuracy criterion for large distances, and make a gravity walk inefficient [4]. Decomposition by equally dividing a space-filling curve overcomes this problem at the cost of a larger surface area between domains due to irregular boundaries. This has the potential of increasing communication needed for more local operations such as nearest neighbor finding. Oct-tree based decomposition yields good aspect ratios, but typically leads to high load imbalance.

The Charm++ language solves the load balancing problem by allowing the work to be divided into many more pieces (chares) than processors. As long as none of these pieces is too a large a fraction of the work, the runtime system can migrate chares among the processors so as to evenly distribute the work. In CHANGA, the tree used for gravity and neighbor finding is split into such pieces, "TreePieces," where each piece is a contiguous set of leaves (or buckets) containing particles and the ancestor nodes of those leaves up to the root of the entire tree. Some of these ancestors have children on other TreePieces, and are referred to as "boundary nodes." The boundary nodes and the TreePieces on which their children reside are discovered during the tree build phase.

Such a data structure can support several types of domain decomposition. The most straightforward decomposition to implement is based on a space-filling curve, such as Morton ordering [212] or Peano-Hilbert curves [226]. In this case, a histogramming algorithm similar to that used in a parallel sort [116, 224] can be used to determine splitting points along the space-filling curve (splitting keys) that divide it into approximately equal numbers of particles. The determination of the splitting keys can be done without any data movement: candidate splitting keys are broadcast to all the processors, which return the contribution of their particles to the domain counts in a reduction. The candidate keys are then adjusted, and the process is iterated until a good set of splitters is determined. Once these splitting keys are specified, the particles can be shuffled among the chares so that each chare contains the particles in its domain. Since the boundaries between nodes on an Oct-tree can also be specified as keys, the same infrastructure can be used to perform a domain decomposition such that each domain corresponds to a complete node in the Oct-tree. In other words, the chares would contain complete nodes, and the union of all the chares would completely tile the computational domain. We refer to this method as "Oct decomposition."

Figure 6.1 shows an example of the regular domains obtained through this procedure. The figure shows the projection on the xy plane of the decomposition of a 100 particle input distribution over a chare array, such that no chare has more than 3 particles. The domains assigned to chares are represented by boxes that have not been split. The reader may notice that certain chares seem to hold more than 3 particles. This is an artifact of the two-dimensional projection of the three-dimensional spatial domain, since splits along the z dimension are not shown in the figure. Also note that we represent an Oct-tree as a binary tree with splits in successively orthogonal directions, so the nodes can be either cubical or have aspect rations of 2:1:1 or 2:2:1. The aspect ratios in the figure differs from these because the domain started as non-square. In the current version of the code, we force the computational domain to be cubical to minimize the aspect ratios of the tree nodes.

The disadvantage of such regular domains is that the number of particles per domain could vary widely. The Charm++ load balancing framework is

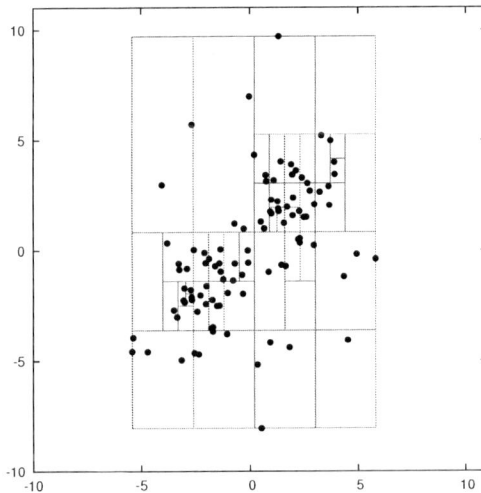

FIGURE 6.1: Decomposition of 100 particles among chares so that no chare has more than three particles. Filled circles represent particles, and straight lines depict the Oct-splitting of domains that hold more than three particles. Domains that are not split at all form the spatial extents assigned to individual chares. Note that splits along the z dimension are not shown.

used to distribute the chares among processors to maintain an even amount of work per processor. The work per chare is determined by measuring the timings from the previous time step, and assuming the principle of persistence, namely that the work done by a TreePiece will change very little between time steps. We have experimented with a number of load balancing strategies, including balancers distributed with the Charm++ system, and balancers specifically designed for the above domain decomposition of particles. The efficacy of some of these load balancers will be demonstrated in some of the benchmarks below.

6.2.2   Tree Building

Once the particles are on the TreePieces, each TreePiece builds its tree independently and in a top-down manner, starting from the root node that contains the entire simulation domain. During this phase, each TreePiece also has information about the extents of the domains of all other TreePieces. This information is used by the tree building algorithm to determine which nodes are *boundary nodes*, i.e., nodes that enclose particles assigned to multiple TreePieces. For example, with more than one TreePiece, the root of the tree is always a boundary node. More formally, if a TreePiece constructs a node that has at least one of its enclosed particles assigned to a *different* TreePiece, then that node is labeled as a boundary node in the local tree constructed by the first TreePiece. The construction of the multipole moments of such a node requires data held by other TreePieces. Therefore, if a child of a boundary node is completely remote, a request is sent to the TreePiece responsible for that child. Finally, note that if a node spans more than one TreePiece (i.e., it is a boundary node in the local tree of each one of the spanned TreePieces), the requests for its moments are distributed over all the owners. Such requests for moment information can arrive at the responsible TreePiece before it has constructed the node in question. In this case, the request is queued on the responsible TreePiece. This queue is processed after the responsible TreePiece has built its own tree, or when its own requests for moments have been satisfied.

Note that as described, every TreePiece would treat only *its own* particles as locally accessible. However, for reasons of efficiency, TreePieces should have direct access to the particles of *all other* TreePieces resident in the *same address space*. For this reason, all the TreePieces on an SMP (symmetric multiprocessor) node are merged. At the end of this merging phase, each TreePiece on an SMP node has access to a shared, read-only tree data structure that represents the union of all the TreePieces in the address space.

Figure 6.2 presents an example of a portion of a distributed Oct-tree that was constructed using the CHaNGa tree building algorithm. This example tree was obtained after Oct-decomposition of a 1000 particle input model over 8 TreePieces. The TreePieces themselves were spread across 4 SMP nodes (address spaces). We show the tree only down to the level of the roots of

1(373)
6:7

2(373)
0:5   3(0)
6:7

4(373)
0:4   5(0)
5:5

8(373)
0:3   9(0)
4:4

16(0)
0:0   17(373)
1:3

34(0)
1:1   35(373)
2:3

70(18)
2:2   71(355)
3:3

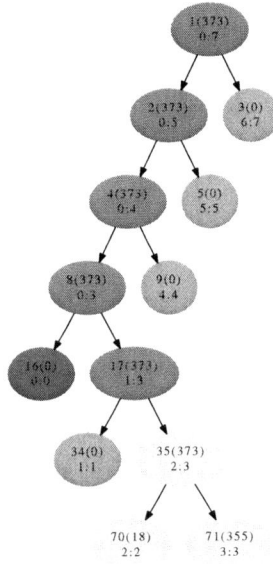

FIGURE 6.2: Portion of the distributed tree constructed after oct-decomposition of 1000 particles over 8 TreePieces. The TreePieces were spread across 4 address spaces. The tree shows the following types of nodes: (i) Boundary (medium gray ovals), (ii) Remote (medium gray circles; node 24 is a Remote Bucket, i.e., a remote leaf node), (iii) Internal (light gray). Each node also shows the key, the number of contained particles in this address space and the range of TreePieces that hold particles contained in the node. The tree is pruned at the roots of individual TreePieces, so that local buckets are not shown.

individual TreePieces. Each node in the tree has three pieces of information: (i) the key of the node, (ii) in parentheses, the number of particles contained by the node *in this address space* and (iii) the range (*first : last*) of TreePieces that hold particles contained by the tree node. The various types of nodes are depicted by their colors. Green nodes are Internal to the address space, i.e., all the particles enclosed by such a node are present in the address space of the tree. In the figure, the nodes with keys 70 and 71 are the roots of the two TreePieces present in this address space. From the ranges specified in these two nodes, we see that TreePiece 2 was assigned all particles enclosed by node 70, and TreePiece 3 was assigned all particles enclosed by node 71. Note that if both children of a node are Internal, then the node itself must be Internal to the address space. This rule is exemplified by node 35, whose children are nodes 70 and 71, both of which are Internal to the address space. Ochre denotes Remote nodes (34, 9, 5 and 3), whereas the lone orange node (key 16) is a Remote Bucket. The special designation of Remote Bucket indicates that it is a leaf node of the tree and remote particles will need to be accessed if this node needs to be opened. Note that none of the particles enclosed by the Remote/Remote Bucket nodes are present in the current address space. Finally, pink denotes Boundary nodes: some, but not all particles enclosed by a Boundary node, are present in the address space. It should be clear that the type of a node is not an inherent property of the node. Instead, it depends on the address space in which the tree was constructed and, in particular, on which TreePieces (and hence, which particles) were present in the address space following decomposition.

6.2.3   Tree Walking

The traversal of the tree to calculate gravity is done by the algorithm described in [230]. The result of this algorithm is similar to the standard Barnes-Hut walk, but it introduces a few optimizations to reduce the number of comparisons needed. Furthermore, it also avoids the large worst-case force errors produced by the standard Barnes-Hut opening criterion [212]. The first optimization is that forces are evaluated on a bucket of particles at a time. While this results in more floating point operations being performed (the forces between particles in a bucket scale as the square of the number of particles in a bucket), the overall performance is improved on modern pipelined architectures. The second optimization recognizes that buckets that are siblings will have very similar interactions with distant nodes. Hence, lists of interactions are generated by walking parent nodes before walking their children. The detailed algorithm is described in [230], but the essence is the following. Given a target node containing particles on which the forces are to be calculated and a list of source nodes which contain particles which provide those forces, each source node is checked for three conditions. First, it is determined if the multipole expansion of the source node is valid for all particles in the target node. A multipole expansion is valid for all space outside a sphere centered on the source node and of a radius determined by the size of the node divided by θ, the standard Barnes-Hut parameter. Hence the first condition is satisfied if this sphere does not intersect the bounding box of the target node. In this case the source node is removed from the list, and its multipole expansion is added to a list that will be computed for all particles in the target node, i.e., a list of interactions that can be reused for all children of the target node. The second condition is whether the bounding box of the target node is completely contained within the sphere. In this case there is no particle in the target node for which the multipole expansion is valid, so the source node will be removed from the list and be replaced by its children, which will in turn be subject to these checks. The third condition is when the bounding box partially intersects the sphere, in which case the source cell is left on the checklist. If the target node is not a leaf, after all source nodes have been checked, then, for each child, the checklist concatenated with the child's siblings is checked with that child as the target.

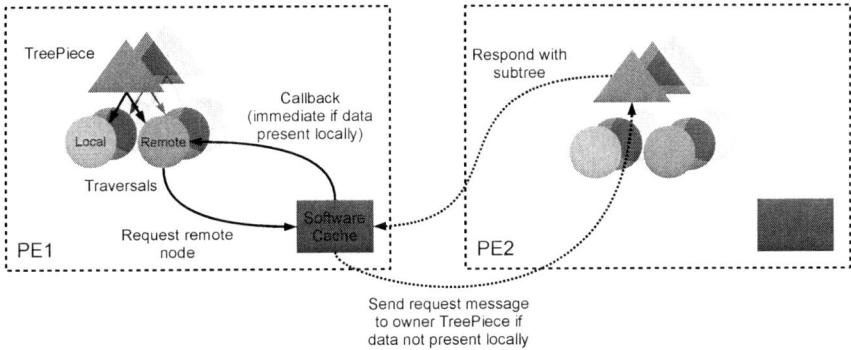

FIGURE 6.3: Flow of control with a software cache for reuse of remotely fetched data.

Performing the above walk nearly always requires a TreePiece to access nodes on other processors. This access is performed through a software cache that is shared with all TreePieces on a processor. The control flow of the cache-augmented code is depicted in Figure 6.3. Upon a cache miss a request is sent to the processor on which the node resides, and the walk causing the miss is suspended, allowing other work to be done, perhaps on other TreePieces. When the cache request is fulfilled, a callback resumes the walk where it left off.

Two optimizations are made to further improve the parallel performance of the code. The first is that a separate "prefetch" walk is performed first to obtain non-local tree node information and reduce cache misses during the main tree-walk. Second, the main tree-walk is itself divided into separate local and remote parts. Walks over the tree local to a processor are performed at lower priority. By giving the remote walk a higher priority, requests for remote data are issued early, giving the opportunity for local computation work to be overlapped with communication latency.

The order of the multipole expansion for evaluating the far field forces due to a mass distribution within a node involves a trade-off. The simplest, fastest and lowest accuracy expansion is the first term of the expansion: the monopole. However, for higher accuracy it becomes more CPU efficient to use higher order expansions. As will be shown below, for the typical force accuracy parameters used in cosmological simulations [201, 202], where the relative force accuracy is better than 1 percent, it is most efficient to keep terms to hexadecapole order.

6.2.4 Force Softening

Force softening is a crucial feature for accurate cosmological simulations; however, multipoles are more cumbersome to compute in the presence of force softening. Because of the symmetries available with a purely Newtonian potential, 30% fewer terms are needed for the moment expansion and a factor of two fewer floating point operations is needed to evaluate the forces at hexadecapole order. Because of these difficulties, we avoid calculating the high order multipole expansions of softened potentials. If a target bucket is within the force softening radius of any of the particles in a source node, the forces are evaluated as a softened monopole. In order to keep the same accuracy, a modified accuracy criterion, $\theta_{Mono} = \theta^4$, is used since errors for a multipole order, p, scale as θ^p.

6.2.5   Periodic Boundary Conditions

Gravitational forces in simulations with periodic boundary conditions are a challenge because of the long range nature of gravity. CHANGA addresses this challenge using the Ewald summation technique [61] where the force is divided into a summation in real space and a summation in Fourier space. The implementation is similar to that of [52] and is described more fully in [230]. For a given particle, non-periodic forces are calculated via the above tree-walk from all particles in the fundamental cube and a number of its replicas, usually the 26 nearest neighbors, but more for higher force accuracy. Then, the Ewald expansion of the multipole moments of the fundamental cube is evaluated for that particle (see [230] for the formulæ) with the real space expansion being modified to avoid double counting the $1/r$ component that was calculated via the tree-walk in the nearby replicas. This technique has two desirable features. First, the intra-particle force law is not modified from the non-periodic force during the tree-walk, keeping it simple and fast, and second, the evaluation of all Ewald expansions requires only a small amount of data, the multipole moments of the root node.

6.2.6   Neighbor Finding

To calculate Smooth Particle Hydrodynamics (SPH) quantities, nearest neighbors need to be found. CHANGA uses a kth nearest neighbor algorithm using the same tree that is constructed for the gravity walk. For each particle, the search is started using k particles that are nearby in tree order. The most distant of these particles defines an initial search radius. The walk then begins by searching the bucket containing the particle, then proceeds up the tree, searching the siblings of the bucket's ancestors. Nodes that do not intersect the current search boundary are skipped. A heap of neighbor particles is maintained during this search, so that after k neighbors have been found, and a new particle is found that is closer than the current furthest particle, the old particle is popped off the heap, the new particle is pushed onto the heap and the search radius is reduced. After the entire tree is searched, the distance to the kth nearest neighbor is stored, and the neighbors are passed to a function to evaluate the SPH quantities.

In general, SPH requires at least two iterations over the particles, once to calculate densities, and a second time to calculate pressure gradients. For the second pass, the neighbor finding is easier since the distance to the kth neighbor is already known. Therefore, a straightforward top-down walk is performed, looking for nodes whose bounding box intersects the sphere defined by the particle and the kth nearest neighbor distance.

6.2.7 Multi-Stepping

The wide range in densities represented in a cosmological simulation also implies a large range in dynamical times. That is, wide variations in mass densities can lead to particle velocities that vary by several orders of magnitude. In such a scenario, it would seem that accuracy can only be maintained by evaluating particle trajectories on the smallest determined time scale. However, we use adaptive time scales, i.e., we evaluate forces and integrate particle trajectories on the dynamical time scales relevant to the particles, rather than evaluating the forces on all particles at the smallest time scale. This method of *multi-stepping* allows us to potentially increase the efficiency of simulation by several orders of magnitude over a corresponding *single-stepped* simulation. Ideally, we would like to evaluate each particle on its individual dynamical time scale. However, there are overheads such as tree building, which must be incurred prior to any gravity calculation step. Several cosmological simulation codes handle this situation by a block hierarchical time-stepping scheme, and CHANGA does likewise. In this scheme, particles are assigned to a time step "rung" which is a power of two subdivisions of the base time step. All particles with a time step on a given rung or smaller are evaluated at the same time, hence amortizing the fixed costs of the gravity calculation.

6.3 Accuracy Tests

6.3.1 Force Errors

To test the accuracy of the tree gravity algorithm we use two particle distributions that are appropriate for cosmological simulations. The first, `lambs`, is a 30,000 particle subsample of a 3 million particle realization of the $z = 0$ state of a ΛCDM (Cold Dark Matter) universe with $\Lambda = 0.7$ and $\Omega_m = 0.3$. The realization is a cube $50\,h^{-1}$ Mpc on a side, where h is the value of the Hubble constant in units of 100 km/s/Mpc. This is a low resolution realization of the simulation analyzed in [202] to study the mass function of dark matter halos. The second, `test4`, is a cosmological zoomed-in simulation where most of the particles are used to focus on the structure of a single object of interest, and the remaining particles are lower resolution and are distributed throughout a cosmological volume to represent the tidal fields of the surrounding structure. In test4 the structure of interest is a roughly Virgo sized galaxy cluster in a 50 h^{-1} Mpc cube with an $\Omega_m = 1$ CDM cosmology. This is the "Dark" simulation analyzed in [155], and contains 67,495 particles. The relatively small number of particles in these test simulations was necessary to compute the exact forces in a reasonable amount of time. We calculated the exact periodic forces using the $O(N^2)$ code direct[1] to compare with the forces calculated by CHANGA. In addition we compared a non-periodic force calculation of test4 as a proxy for an isolated cluster.

The hexadecapole expansion used in CHANGA is a higher order than many tree codes. Higher order is proven to be more efficient for higher accuracy forces in the FMM algorithm [89], and presumably this is the case for the Barnes-Hut algorithm as well. We investigate this by showing the trade-off between force accuracy and computational cost in Figure 6.4 for quadrupole (dotted lines) and hexadecapole (solid lines) versions of CHANGA. The wall-clock time on a single core of a 2.33 GHz Intel Core2 Duo processor for a force calculation is plotted against the accuracy of the calculation for the three test problems. Accuracy is quantified as the RMS relative force error over all particles. For each of the simulations we varied θ from .3 to 1.0 to obtain the range in computational cost and force accuracy.

By comparing the dotted with the solid lines, we see that indeed for higher force accuracies, the hexadecapole achieves a given force accuracy at a lower computational cost, while at the lowest accuracies, the quadrupole is faster. For the simulations with periodic forces the crossover is at about 1 percent. That is, if a cosmological simulation requires better than 1 percent RMS force accuracy, then the hexadecapole will provide the fastest force calculations. [202] showed that simulations with PKDGRAV using $\theta = .7$ (giving an RMS force error of about 1 percent) gave converging results for halo mass functions. Force errors can also influence the structure of collapsed objects, in particular leading to the development of artificially low density cores [201]. Although [201] used a force accuracy criterion that resulted in median relative force errors of about 0.001, they point out that even a few very large force errors can lead to deviations from their converged results. In Figure 6.5 we show the maximum error for the same parameters as in Figure 6.4. For the cosmological simulations, the computation required for a given maximum force error is almost always less for the hexadecapole than for quadrupole; hence expanding to hexadecapole order is a good choice for these simulations.

6.3.2   Cosmology Tests

In this subsection we test the ability of CHANGA to reproduce cosmological test problems that have been published in the literature. To test the gravity calculation in a cosmological context we look at the test problems used

---

[1]See http://www-hpcc.astro.washington.edu/tools/direct.html.

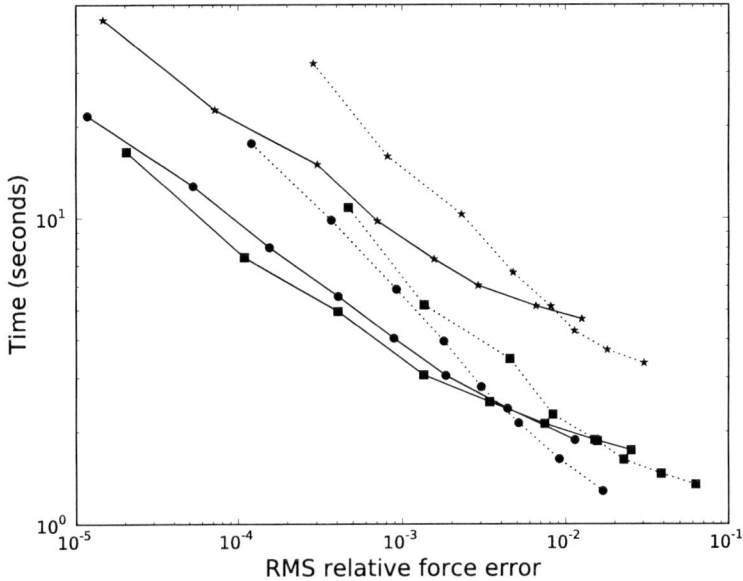

FIGURE 6.4: Cost of force calculations as a function of force accuracy. The wall-clock time for a force calculation is plotted against the RMS relative force accuracy averaged over all particles. The solid lines show the results for the hexadecapole expansion while the dotted lines show the results for the quadrupole expansion. Square points indicate the cosmological volume (lambs), starred points indicate the cluster (test4) and circular points indicate the cluster without periodic boundary conditions. On each line, points are plotted at $\theta = .3, .4, .5, .6, .7, .8, .9, 1.0$.

in [98] that follow the large scale structure formation in a uniform periodic volume of only dark matter. For a combined test of gas dynamics and gravity, we use a test problem introduced by [60]: the spherical collapse of adiabatic gas. This problem is an idealized version of a galaxy formation simulation with a high Mach number shock.

Reference [98] compared a number of simulation codes with regard to producing results for large scale structure. They also provided their initial condition data for others to make comparisons. One of the more important comparisons is the mass function of halos since this is a simulation result that is widely used to compare with observations of large scale structures. In Figure 6.6 we compare the results of ChaNGa with the published results from GADGET for the mass function from their *LCDMb* simulation: a 360 Mpc simulation with 256^3 particles of a ΛCDM universe. As in [98], halos are found using a friends-of-friends (FOF) group finder with a linking length of 0.2 times

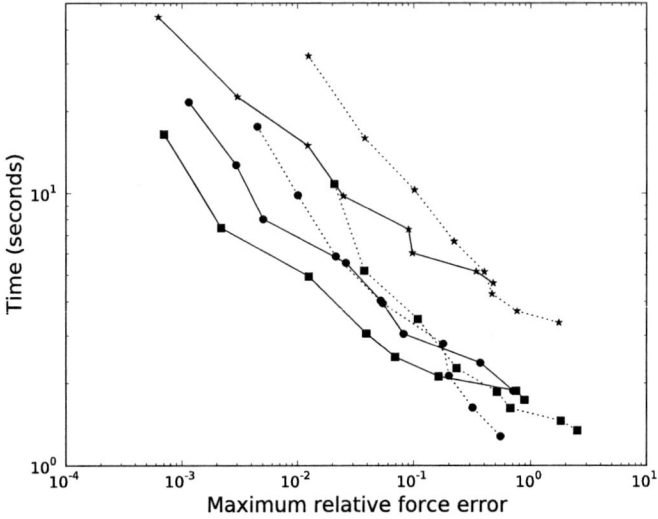

FIGURE 6.5: Cost of force calculations as a function of maximum error. The wall-clock time for a force calculation is plotted against the maximum relative force error. Lines and points are as in Figure 6.4.

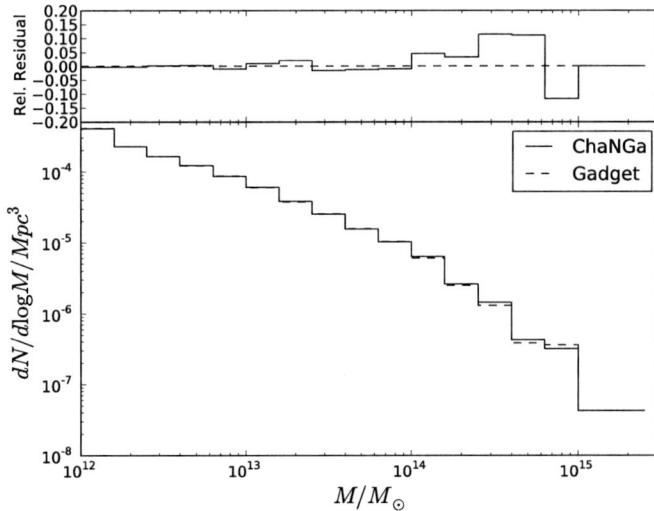

FIGURE 6.6: The halo mass function of a 360 Mpc simulation. The FOF halo mass function from the [98] 360 Mpc cosmological simulation run with CHANGA (solid histogram) is compared with that derived from the published GADGET data (dashed histogram). The upper panel shows the relative difference compared to GADGET.

the mean inter-particle separation. The halos are placed in logarithmic mass bins, and the number density of halos per logarithmic mass interval is plotted as a function of mass. Comparing Figure 6.6 with Figure 26 of [98], we see that the difference between the CHANGA and GADGET results is comparable to or less than the differences seen between the other codes tested in [98]. This gives us confidence that CHANGA is useful for making predictions of the large scale structure of dark matter.

Another often used result of cosmological N-body simulations is the density profile of collapsed objects. The Santa Barbara Cluster Comparison Project [69] compared a number of simulation codes in simulating the formation of a cluster of galaxies in a CDM universe. This simulation has also been used by [98] to compare dark matter-only simulations. In Figure 6.7, we compare the cluster at $z = 0$ for a 256^3 particle simulation run with CHANGA (solid line) and with GADGET (dashed line). The CHANGA simulation was run with a comoving spline softening length of 10 kpc, while the GADGET simulation is reported to be run with a softening of 50 kpc, Plummer equivalent. The profiles

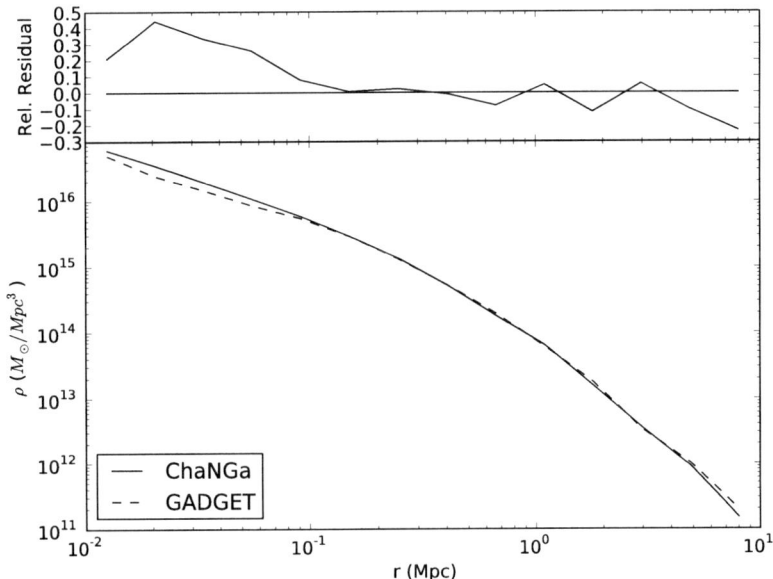

FIGURE 6.7: The density profile of the Santa Barbara cluster. The dark matter density averaged in logarithmically spaced spherical shells is plotted as a function of radius for the Santa Barbara cluster, 256^3 particle simulations. The GADGET results are those published by [98]. The upper panel shows the relative difference compared to GADGET.

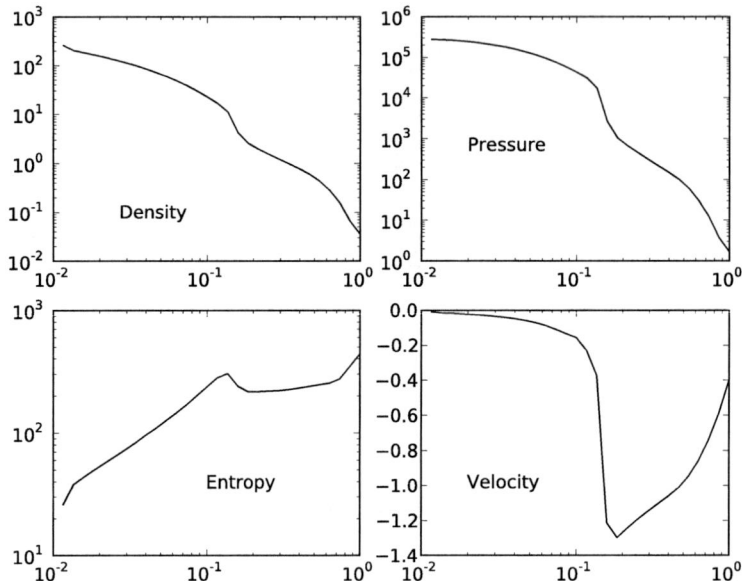

FIGURE 6.8: Adiabatic collapse test from [60] with 28,000 particles shown at time $t = 0.774$. The results from CHANGA are shown overplotted on the Gasoline results. The differences between the two codes are less than .5% and are not visible on this plot.

agree quite well down to the resolution limit of the GADGET simulation: a couple of softening lengths or 100 kpc.

Finally, we present a test of the SPH implementation. A useful simplification of galaxy formation is the collapse of an initially cold self-gravitating sphere of gas. Reference [60] used a test of this scenario that we duplicate here. In this test most of the cold gas eventually gets heated through a shock that at one point has a Mach number greater than 2. In Figure 6.8, we show the profiles of the hydrodynamic variables midway through this collapse. The initial conditions were realized in a three-dimensional "glass" of particles as in [240]. For the comparison we averaged the hydrodynamic quantities in logarithmically spaced spherical shells. The tight concordance between Gasoline and CHANGA should come as no surprise, since the same formulation of SPH is used in both codes.

6.4   Performance

For performance evaluation at scale we introduce several datasets larger than those used in the accuracy tests above. First, we have a low resolution (144^3 particles) and a high resolution (432^3 particles) realization of a ΛCDM universe with $\Lambda = 0.7$ and $\Omega_m = 0.3$. These simulations are from [202], and are referred to as "3M" and "80M" for the small and large dataset, respectively. We also use an intermediate resolution data set (256^3 particles) which is the *LCDMb* simulation from [98]. We refer to this dataset as "16M." All these datasets are uniform in the sense that they are the final outputs of a simulation that started with equal mass particles uniformly sampling a periodic volume at high redshift and evolved to the present.

It is a common practice in the study of galaxy formation to perform "zoom-in" simulations where a galaxy of interest is realized with a large number of high resolution particles, while the remainder of a cosmological volume is represented with low resolution particles. These kinds of simulations present a challenge to domain decomposition since most of the computation is done in a small spatial volume. This issue is exacerbated by the fact that the dynamical time within the galaxy is much shorter than the overall dynamical time of the Universe. Hence the forces on particles within the high resolution region need to be evaluated much more often than on the particles in the rest of the volume. For the performance tests below we have two such datasets. Both are realizations of a dwarf galaxy forming within a 25 Mpc cube. These are dark matter only versions of the "Dwarf" galaxy formation simulation studied in [86]. The low resolution (5 million particles) and high resolution (50 million particles) are referred to as "5M" and "50M," respectively.

6.4.1   Domain Decomposition and Tree Build Performance

To begin with, we discuss the performance of the tasks that support the eventual traversal of the distributed tree to calculate gravitational forces. The scaling performance of the force computation phase itself will be discussed subsequently. As described in Section 6.2, every iteration begins with the decomposition of particles into TreePieces. In order to gain a better understanding of the performance of the code, we further divide the decomposition phase into a *histogramming* sub-phase and an *all-to-all* sub-phase. The histogramming phase consists of a number of iterations to determine a good set of splitters for the domains. At the end of this phase, each TreePiece knows the range of particle keys held by every other TreePiece. Using this information, particles are flushed to their respective owners in an *all-to-all* operation. This is followed by the distributed construction of the tree, the algorithm for which was discussed in Section 6.2.

Figure 6.9 shows the time taken to complete these tasks for the 50M data-

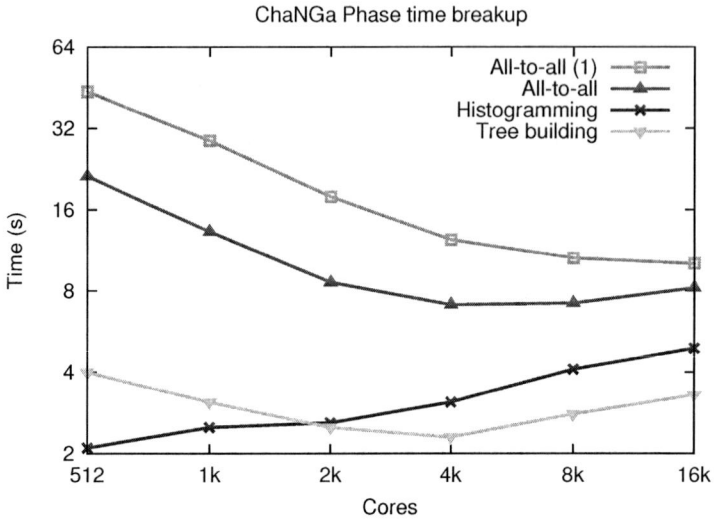

FIGURE 6.9: A breakdown of time taken by different tasks in CHANGA. The graph shows time taken to obtain an Oct decomposition of particles onto TreePieces via a parallel, iterative histogramming technique (Histogramming); time to shuffle particles between TreePieces once the decomposition has been ascertained (All-to-all) and the time to construct the distributed tree data structure (Tree-building). A separate line is shown for the first all-to-all in the simulation.

set for a wide range of core counts. First, note that the time for histogramming *increases* gradually, instead of decreasing, with an increase in core count. This is because of the sequential bottleneck at the master chare that evaluates the splitters and adjusts them, as well as the increase in communication time due to larger configurations of processors. Next, observe that the time to construct the tree in parallel initially scales down with increasing core counts, but stops scaling at 4k cores and increases thereafter, ostensibly due to the large amount of communication that must be performed at 8k and 16k cores. Finally, we show two curves for the all-to-all phase. The first of these, labeled "All-to-all (1)," depicts the time taken to complete the all-to-all exchange of particles in the *first* decomposition step. The curve labeled "All-to-all" shows the average over ten iterations of subsequent instances of the decomposition phase. The reason that the first all-to-all takes significantly longer than the following ones is that for the initial decomposition, particles are read out of input files that are not in tree order, so that each TreePiece initially holds an arbitrary subset of particles. Therefore, during the first decomposition, a large amount of data must be exchanged between TreePieces. In subsequent iterations, since particles move relatively small distances, the majority of the particles are assigned to the same TreePieces that they were in the previous

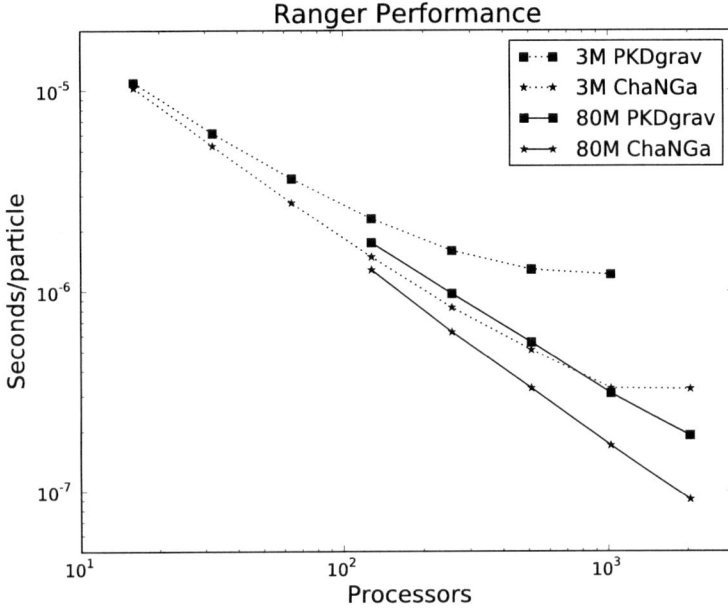

FIGURE 6.10: The time to complete a single gravity calculation for all particles is shown as a function of the number of cores used on the TACC machine "Ranger." Results for two uniform volume datasets are shown, one with three million particles and one with 80 million particles.

iteration. Consequently, less data is exchanged between processors in the all-to-all phase.

6.4.2   Single-Stepping Performance

In Figure 6.10, we compare the performance of CHANGA with the legacy code, PKDGRAV, running on Ranger, an Infiniband connected cluster with 16 cores per node. Strong scaling results are shown for the cosmological datasets 3M and 80M. The time taken to calculate the forces on all particles is plotted as a function of cores used; therefore, perfect scaling would be a line of slope -1. For the 3M dataset, the PKDGRAV scaling flattens out after 256 cores, while CHANGA continues to scale up to 1024 cores, at which point there are less than 3000 particles per core.

To evaluate the performance of CHANGA on larger core counts, we ran tests on Intrepid, an IBM Blue Gene/P. Figure 6.11 plots the core-hours taken for a gravity calculation on all particles as a function of core count. In this figure, a horizontal line indicates perfect scaling, while the dotted lines with positive slope shows where the wall-clock time taken does not improve

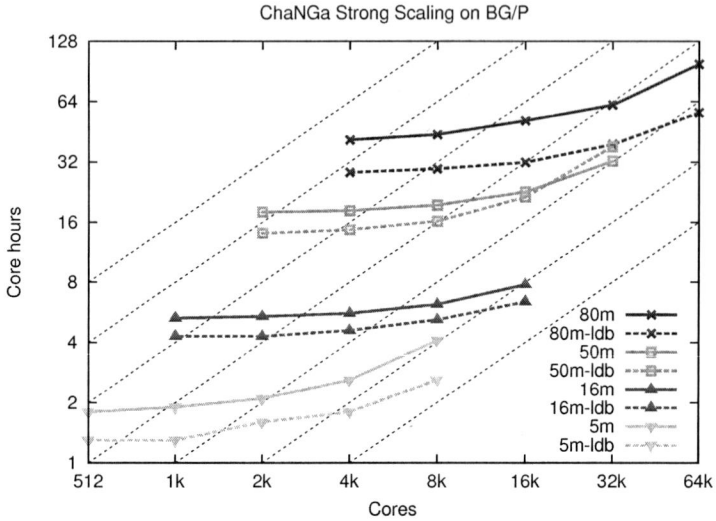

FIGURE 6.11: The core-hours for computing forces on all particles are plotted as a function of the number of cores used. No load balancing strategy was employed for the solid curves. Dashed lines represent scaling performance with the Orb3dLB strategy. The dotted diagonal lines contain points with equal execution time: the code ceases to scale as the slope of a curve approaches that of one of the dotted diagonal lines.

with increasing core count. No load balancing was done for the benchmarks indicated by solid lines, and yet, CHANGA scales to the point where there are a few thousand particles per core.

The dashed lines in Figure 6.11 show the wall-clock time taken to perform the gravity calculation on all particles as a function of core count after load balancing. The load balancer employed here was designed for a tree-based gravity calculation. It obtains the centroids of the particles in each TreePiece, and uses orthogonal recursive bisection in the three spatial dimensions (Orb3D) to divide the TreePieces such that each processor has equal work. In this way, the load is balanced while preserving spatial locality among the TreePieces on a given processor. For the 16M dataset, the load balancer gives a ∼ 20% improvement in performance on 2048 cores through 16,384 cores.

The zoom-in simulations can gain more from load balancing because the cost of force evaluations can vary widely from low to high resolution regions. The advantage of the Orb3D load balancer for this case is shown by the curve labeled "5m." For this dataset, the load balancer gives a ∼ 40% improvement in performance on 1024 cores, while giving nearly a factor of two performance improvement on 8192 cores. The gains in efficiency are similarly impressive for the 80 million particle dataset: on 4k cores, we see that load balancing yields a gain of 1.44 times, with the gain increasing to 1.74 times on 64k cores.

6.4.3 Multi-Stepping Performance

We now evaluate the performance of CHANGA on the multi-stepping challenges presented by the zoom-in simulations. As well as the challenge of computational hotspots, these simulations also violate the Charm++ principle of persistence. This is because of the interleaved nature of the hierarchical time-stepping scheme. For example, if there are particles on the base time step (rung 0), half the base time step (rung 1) and one quarter the base time step (rung 2), force evaluations happen in the following sequence. First all particles get their forces evaluated, then the particles on rung 2, then the particles on rung 1 and 2, then the particles on rung 2 and finally all the particles again. Furthermore, the particles on rung 2 are likely located in a dense region at the center of the simulation. Hence, the number and location of the particles having their forces evaluated changes drastically from time step to time step. To address this problem, we have developed a multi-step load balancer which records the force evaluation loads independently for each level in the time step hierarchy. During the load balancing stage, the loads from the previous time the forces were evaluated at the same rung are used to balance the TreePieces across processors for the upcoming force calculation.

Strong scaling results for this strategy on the TACC Ranger cluster are shown in Figure 6.12. The performance for the 5M simulation is based on the wall-clock time taken to advance the simulation 40 million years of cosmic time for the 5M simulation or 20 million years of cosmic time for the 50M simulation. The 5M simulation has particles distributed over 7 different time step rungs, with the smallest time step being .6 million years. The 50M simulation also has particles distributed over 7 different time step rungs and the smallest time step is .3 million years. The higher resolution simulation requires shorter time steps because of the higher densities. Although the scaling does not go out to as many processors as the single step results, this is a much harder scaling problem: in the 5M run, half the particles only have their force evaluated once over 64 small time steps, while only 130 thousand particles are having their force evaluated every small time step.

The strategy that PKDGRAV uses to load balance across multiple time steps is somewhat different. At each small time step, particles on which forces have to be calculated are balanced across all processors by cost, while inactive particles are balanced across all processors by number. This introduces a non-locality in the data that may additionally impact the scaling of PKDGRAV.

6.4.4 ChaNGa on GPUs

The advent of general purpose computing on graphics processing units (GPGPU) presents an opportunity and a challenge for floating point intensive

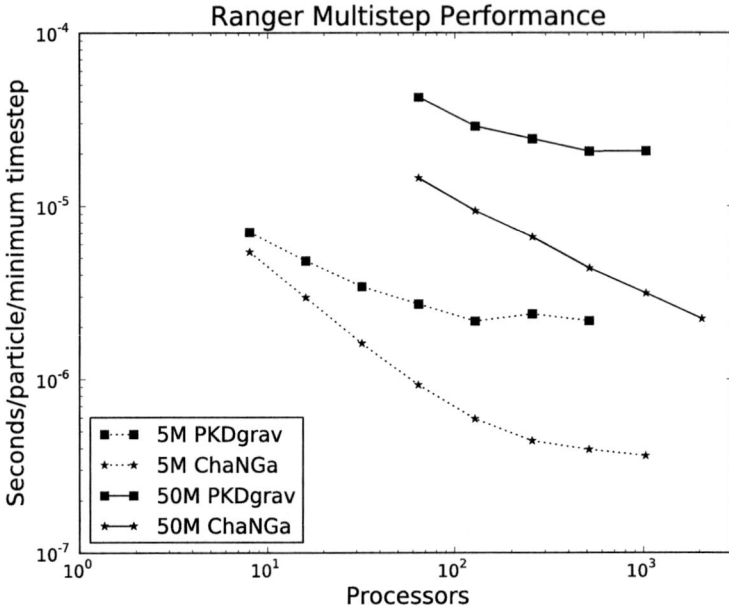

FIGURE 6.12: Multi-stepping performance comparison of CHANGA versus the legacy code PKDGRAV on the Ranger Infiniband cluster. The mean time to advance one particle on the minimum time step is plotted as a function of the number of processors. Ideal scaling is a line with slope -1. The zoom-in data sets 5M and 50M are used for this comparison.

calculations. The opportunity is the remarkable floating point performance of the graphics devices, both in terms of floating point performance/cost and floating point performance/power consumption. For this reason, many of the capability class computing facilities being deployed (e.g., NCSA Blue Waters and ORNL Titan) involve some amount of GPGPU hardware. The challenge is that the performance is achieved through a SIMD architecture and the expectation of a very uniform memory access pattern. A tree code can be particularly difficult to adapt to this kind of architecture because of its irregular data access patterns. A further challenge is that the data for the GPU's computation must be explicitly transferred to the device. This transfer must be orchestrated in a way that does not block computation on either the GPU or the CPU. In Section 6.4.4.1 and afterward, we describe the adaptations we have made to CHANGA to address these challenges, and then demonstrate the performance of the GPU enabled code. We have used NVIDIA's CUDA platform and programming model to implement these changes.

Before discussing how CHANGA has been tuned for GPU clusters, we provide a brief, high-level view of the code in this context. This will aid the

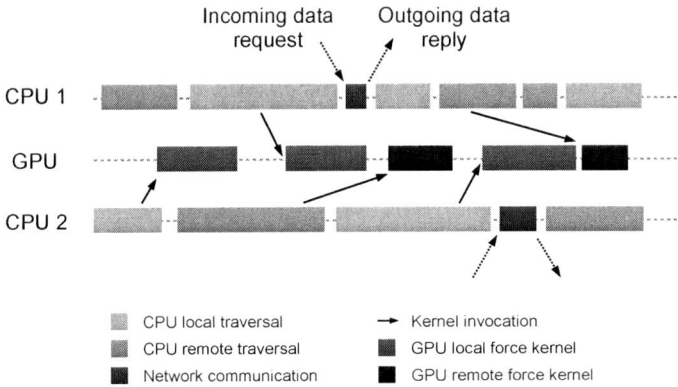

FIGURE 6.13: Schematic representation of the GPU-enhanced version of CHaNGA. Decomposition and tree building (not shown) and tree traversal are done in parallel on the CPUs. Tree traversal results in the construction of *interaction lists*, which are transferred (solid arrows) to the GPU, resulting in force computation kernels.

discussions in Section 6.4.4.1. The GPU-enhanced version of CHaNGA is hybridized, in that some activities are performed on the CPU, and others on the GPU. Figure 6.13 shows some of the elements of the partitioning of activities between the CPU and GPU. In particular, domain decomposition and distributed tree construction (not shown in the figure) are performed on the CPU, since these are memory- and network-intensive operations. The traversal of the tree structure, too, is performed on the CPU, and results in the construction of *interaction lists*. This list construction activity is depicted in the schematic in Figure 6.13 as bands of green (local traversal) and orange (remote traversal). Constructed lists are then asynchronously transported to the GPU for computation of gravitational forces via GPU kernels (blue and violet bands). Forces are then communicated back to the CPU, where they are used to integrate the trajectories of particles. Finally, computation of long-range Ewald forces is offloaded to the GPU, although we do not discuss the optimizations for those kernels here.

6.4.4.1 Adaptations for GPU Architectures

Arranging computation in the kernel. Force computation accounts for a significant portion of the execution time of CHaNGA. Therefore, it is a prime candidate to be moved to the GPU. The force kernel takes as input a list of tuples of the form (b, l_b), where each b is a bucket of *target* particles on which forces are to be computed, and l_b is an *interaction list* of force *sources* for b. These sources can either be particles or tree nodes. A *block* of threads is launched for every such (b, l_b) pair. Although blocks are allowed to be linear in CUDA, we designed them to be two-dimensional, so as to minimize the number of memory operations issued and maximize the amount of concurrency available. Each row of a block corresponds to a target particle, and each column corresponds to an interaction. The dimensions of these blocks are obtained empirically, and the best performance is achieved with 16 rows and 8 columns, for a total of 128 threads per block [109].

Traversal-computation tradeoff. Consider the division of labor between the CPU and the GPU. Whereas the bulk of the work (in terms of number of operations) is performed on the GPU in the form of force computations, the interaction lists upon which the GPU operates must be constructed on the CPU via traversals of the distributed tree data structure. We prevent a CPU bottleneck from forming by increasing the maximum number of particles allowed per bucket. This creates shallower trees and therefore reduces the time spent in tree traversal on the CPU. On the other hand, this allows the bounding boxes of buckets to grow, thereby forcing an increase in the number of particle-particle interactions performed. Figure 6.14 demonstrates this effect for a small ~110K particle dataset running on one CPU and one GPU.

We explain the blue curve in Figure 6.14 as follows: as the maximum bucket size increases, the average depth of constructed trees decreases, thereby causing a commensurate decrease in the number of particle-node interactions. At the same time the number of particle-particle interactions (green curve) rises rapidly under the combined effect of two factors: (1) the increased size of bucket bounding boxes, which causes more nodes to be opened for every bucket for which the tree is traversed, and (2) the occurrence of buckets at

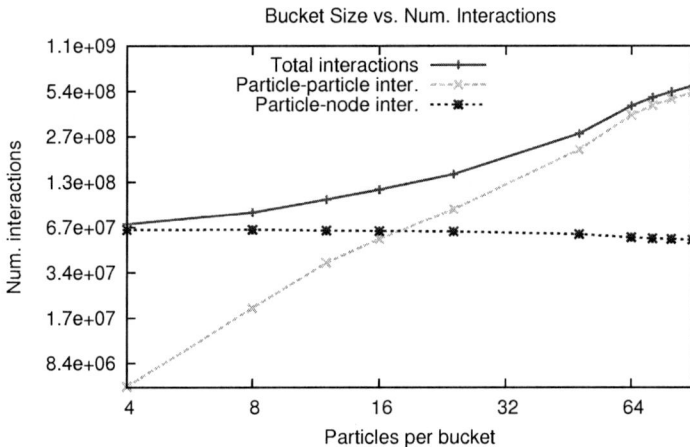

FIGURE 6.14: The total number of interactions increases with an increase in the maximum bucket size.

Traversal vs. force computation tradeoff

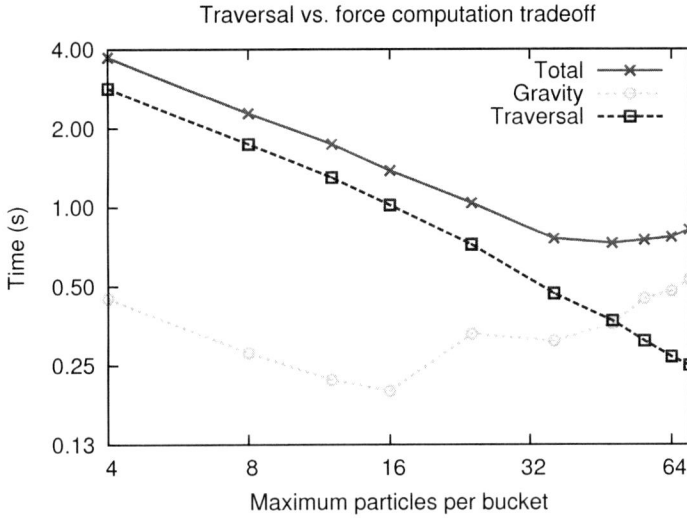

FIGURE 6.15: Traversal time (CPU) decreases and force computation time (GPU) increases with larger buckets, yielding an optimal value for maximum bucket size. Data obtained from a one CPU, one GPU run using the 110k particle dataset.

shallower depths in the tree, so that the opening of a node is more likely to cause particle-particle interactions. This causes a net increase in the total number of interactions (red curve).

Figure 6.15 shows the traversal-computation tradeoff that is engendered by increasing bucket size. The blue curve shows the time taken to perform tree traversal on the CPU whereas the green curve shows the time taken to compute forces on the GPU. Finally, the red curve shows the total of these two activities. Note that the computation times include overheads, such as data transfer to and from the GPU, and kernel invocation overheads. As can be seen, traversal time dominates the total for small bucket sizes. Moreover, traversal time decreases rapidly as bucket size is increased. The force computation time bears a less straightforward relation to the bucket size. Recall that the interactions of a single bucket are assigned to a single block of threads. As the number of buckets decreases, the total number of kernel invocations decreases.

Furthermore, the efficiency of these kernel invocations increases, as explained below. Consider a rectangular iteration space $S = l_b \times w_b$ of a bucket of particles, where l_b is the number of source interactions for the bucket and w_b is the number of target particles in the bucket, and the tile $\tau = l \times w$ is used within a doubly nested *for* loop that covers S. If l_b is not a multiple of l or w_b is not a multiple of w, the block of threads assigned to b will suffer idling threads and control flow divergence. The fraction of instances of the doubly nested *for* loop for which this will happen is about $(\lceil l_b/l \rceil + \lceil w_b/w \rceil)/\lceil l_b/l \rceil \lceil w_b/w \rceil$. As is evident, the denominator of this expression grows more rapidly than the numerator when l_b and w_b (both of which increase with greater bucket size) increase. Therefore, the efficiency of force computation kernels grows as bucket size increases, causing the initial fall in force computation time.

However, as bucket size is increased further the time taken to perform extra computation overcomes the benefits of improved kernel efficiency and the force computation time begins to rise. The point at which this increase causes the red curve to assume an upward trend is the optimal maximum bucket size, $B_* = 48$. This value is fairly constant across datasets and GPU/CPU counts. As is evident from Figure 6.15, the choice of maximum bucket size has a significant impact on performance—choosing a value of 48 improves performance by up to 4 times over a value of 4.

Overlapping CPU and GPU work. In each iteration of a simulation, interaction lists are constructed on the CPU and transferred to the GPU for force computation. In order to maximize GPU occupancy and CPU-GPU transfer bandwidth utilization, interactions are streamed through the GPU. This sets up a two-stage pipeline, whose stages are list construction and force computation. The number of interactions processed in each instance of either stage is determined by the work request (WR) size. Figure 6.16 demonstrates the effect of work request size on execution time.

The data for Figure 6.16 was obtained from the one CPU, one GPU configuration simulating the motion of 110k particles. There are two sources of overhead associated with the hybrid CPU-GPU approach to computing forces. The first is incurred on the CPU, in compiling and copying interaction lists to dedicated buffers that support asynchronous transfers to the GPU. The second is the transfer of the lists themselves, and the subsequent invocation of GPU kernels which receive them as input. A smaller WR size increases the copying and transfer overheads, but at the same time allows the program to exploit the asynchronous memory transfer feature of CUDA to overlap CPU work with GPU activity. A larger WR size creates fewer and more efficient instances of the pipeline, but limits the amount of concurrency exposed. Even though the optimal value of the work request size is largely independent of the dataset used (experiments not shown), it does depend on the number of CPUs and GPUs in use. In order to dynamically tune this parameter, one could apply the Charm++*control points* framework [54]. Lastly, although the value chosen for this parameter can have a discernible effect on the performance, it is still only in the region of 10-20%, and diminishes with core count in strong scaling experiments [109].

Specialized memory management. In order to utilize the asynchronous memory transfer feature of CUDA, one must set up dedicated buffers of memory on the CPU by *pinning* the associated virtual pages to physical memory. In Charm++, this provision is made by the GPUManager API. Internally, the

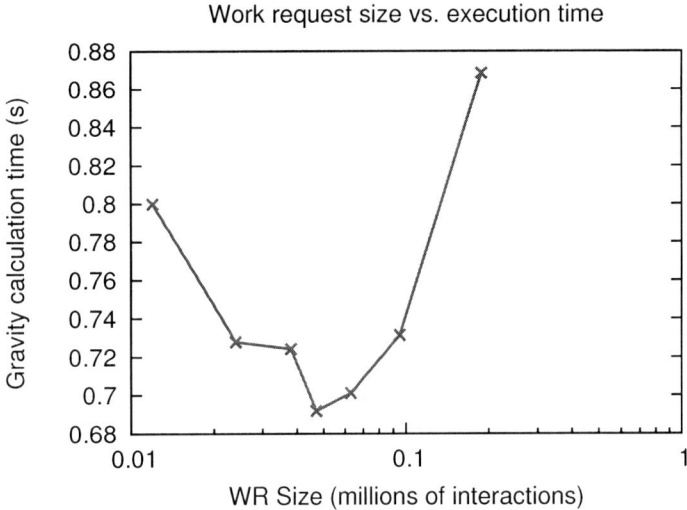

FIGURE 6.16: Optimal WR size for streaming interactions to the GPU. A larger WR size corresponds to less transfer overhead and less concurrency, whereas a smaller WR size leads to more transfers but greater overlap between the CPU and GPU.

GPUManager creates a pre-allocated pool of memory buffers. The individual buffers in the pool are registered with the CUDA runtime in order to pin them to physical memory. By creating such a pool of buffers the GPUManager allows them to be recycled by user code, thereby avoiding the overheads of repeated allocation and pinning.

However, the performance benefits of this memory-pool approach extend further. It was observed in [109] that the CUDA runtime can stall the CPU when it attempts to free buffers of pinned memory allocated on the CPU. Ostensibly, this happens because the CUDA runtime must wait while the GPU is busy executing a kernel. As shown in the context of CHANGA in Figure 6.17, this can add significant overhead to the execution at scale.

Figure 6.17 shows the performance of a 3 million particle simulation on a range of CPU/GPU counts. The green curve shows that performance suffers when no pooling scheme is used and memory management is left to the CUDA runtime system. The red curve shows the performance obtained when buffers are recycled via a memory-pooling scheme provided by the GPUManager API. Note that with this improvement, the performance is close to the upper bound of traversal time, which indicates that the GPU use overheads are extremely low.

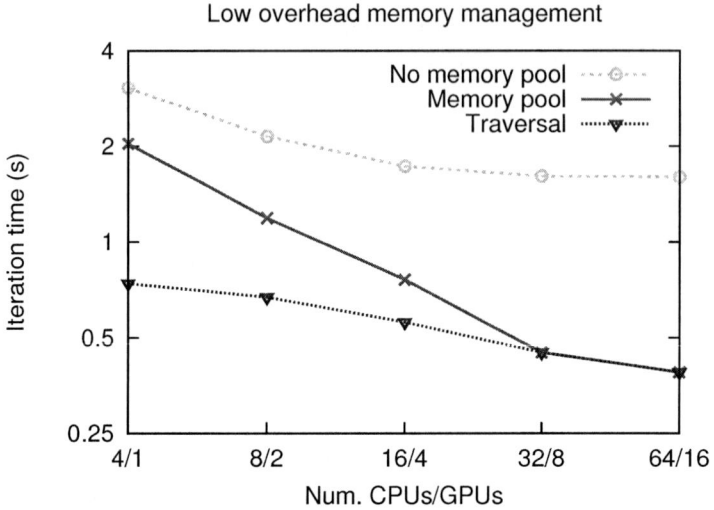

FIGURE 6.17: Recycling asynchronously transferred buffers through a memory pool can yield significant savings in execution time.

6.4.4.2   Performance

We now provide a brief overview of the performance behavior of the GPU-enhanced version of CHANGA. The strong-scaling results presented in Figure 6.18 were obtained from the 3M, 16M and 80M datasets. In contrast to other work [96, 143] on GPU-augmented N-body codes, we used small datasets to demonstrate that reasonable speedups over CPU versions can be achieved with extensive tuning.

The dashed lines in Figure 6.18 show the strong scaling profile of the CPU-only version of CHANGA whereas solid lines show performance with GPU augmentation. These performance figures were obtained on the NCSA's *Lincoln* supercomputer. This cluster has an Infiniband interconnect, and each node of the cluster comprises an Intel Core 2 Quad Q6600 processor running at 2.4 GHz, augmented with an NVIDIA GeForce 8800 GTS card with 128 streaming processors. In our experiments, the CPU-only version of CHANGA used all 8 cores available on each multicore node, whereas the GPU-enhanced version used only 7 of the 8 available cores. For the latter version of the code, it is beneficial to leave one core "free," ostensibly to host the CUDA runtime and other OS processes. Therefore, in the following paragraph when performance numbers are presented for configurations of the form "x nodes/y GPUs," it is meant that the CPU-only version of CHANGA used $8x$ cores, whereas the GPU-enhanced one used $7x$ cores and y GPUs. Moreover, since there are 2 GPUs per node, we have $2x = y$.

Lines in deep red, orange and yellow show performance using the 80M,

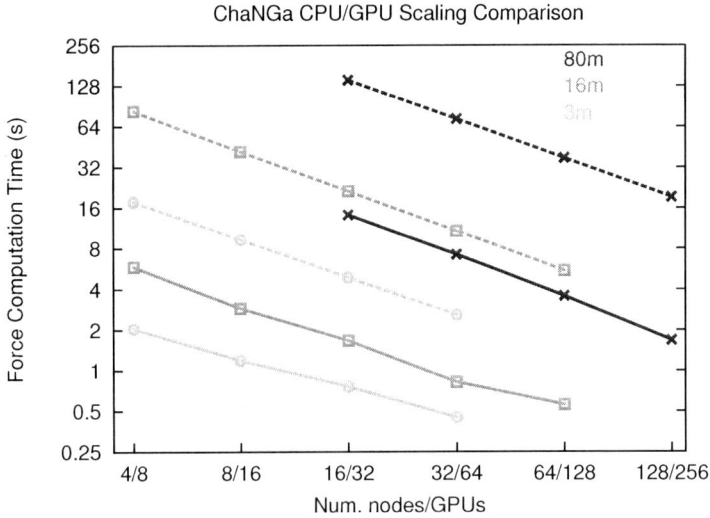

ChaNGa CPU/GPU Scaling Comparison

FIGURE 6.18: GPUs enhance the performance of CHANGA significantly. Dashed lines show scaling performance of CHANGA on multicore nodes. Solid lines show performance with the addition of GPU accelerators.

16M and 3M data sets, respectively. Both the CPU-only and GPU-enhanced versions exhibit good scaling with the 16M and 80M data sets. For the 80M data set, the addition of GPUs yields a speedup of 9.9 on 16 nodes/32 GPUs and 10.5 on 128 nodes/256 GPUs. At this scale, the simulation achieves an average computation rate of 3.82 TF/s. Similarly good results are achieved with the 16M data set. The GPU-enhanced version of CHANGA is 14.1 times as fast as the CPU-only version on 2 nodes/4 GPUs. This speedup is sustained until 32 nodes/64 GPUs, where the GPU-enhanced version is 13.2 times as fast as the CPU-only version of CHANGA. Efficiency drops slightly at 64 nodes/128 GPUs, yielding a speedup of 9.8 at that point. On the 3M data set, the GPU-enhanced version of CHANGA is 9.5 times as fast as the CPU-only version when run on 2 nodes/4 GPUs. This speedup falls gradually to 5.8 on 32 nodes/64 GPUs. Although the code does scale to 64 nodes/128 GPUs with the 3M data set, the improvement in time is marginal from the previous data point, and is therefore not included in Figure 6.18.

6.5 Conclusions and Future Work

For over a decade collaborators using the MPI based N-body/SPH code, PKDGRAV/GASOLINE, have published high impact scientific results ranging from cosmology (e.g., [177, 85]) to planet formation (e.g., [164]). Hence, one should question why an effort was made to develop a new code based on Charm++ rather than devoting that same effort to improving the performance of the existing code. Some of the reasons are as follows. First, the difficulty in load balancing a multi-stepping simulation was obvious, so an established framework that could help with that task was desirable. Also, the MPI code did not include a mechanism for overlapping computing and communication. Instead, communication costs were amortized using the software cache. This strategy became less and less viable as processor speeds increased faster than network latencies decreased. Furthermore, SPH algorithms do not generate enough floating point work compared with the relatively more costly communications. Finally, large machines were becoming more hierarchical. Effective use of these machines would at least require combining on-node threading with the MPI code. Learning a new parallel programming system did not seem that much more daunting than developing a threading + MPI code.

Was this effort worth it? While it is hard to answer the hypothetical question of whether the same amount of effort would have improved the MPI code, what we can say is that the results above show that the effort was successful in producing a more scalable, scientifically usable code. At the 1000 processor level, CHANGA increases the overall throughput of a realistic cosmological simulation by a factor of about 5 over the legacy code. Meanwhile, it demonstrates strong scaling out to 10s of thousands of cores. Furthermore, an initially unexpected result was the ability to port CHANGA to GPU clusters, resulting in a 10-fold speedup in the gravity calculation.

There are a number of Charm++ features that enabled the success of CHANGA.

- Computation/communication overlap. Not only by dividing the data into TreePieces, but also by dividing the gravity calculation into separate remote walks, local walks and Ewald calculations, we took full advantage of having computation to overlap any communication.

- Entry method prioritization. By giving entry methods that respond to data requests higher priority, we ensured data was available for further computation.

- Load balancing framework. The dynamic nature of the tree calculation makes *ab initio* determinations of the load difficult, so measurement-based load balancing is a sensible strategy. However, a flexible framework was needed to make good use of the load information, as in the multi-step load balancing scheme.

- Composability. Independent parts of the calculation, such as gravity and SPH, could be performed concurrently. This gives further opportunity for overlapping computation and communication, as well as making load balancing easier.

- C/C++ based. Many of the basic routines for calculating gravity and SPH forces in CHANGA were taken directly from PKDGRAV/GASO-LINE. The ability to reuse existing well-tested code greatly reduced the effort needed to develop CHANGA.

Of course, Charm++ was not a "silver bullet" for the development of a parallel N-body code. In fact, our first design, which did not include the software cache, had extremely poor parallel performance. Despite the features of Charm++, it is obvious that one must still think carefully about data movement in a parallel application.

There are also a number of improvements to be made. First, a better gravity algorithm, the fast multipole method, has been implemented in PKDGRAV2[229], and so can be easily ported to CHANGA. Secondly, multiple time steps continue to provide challenges that should be addressed. Avenues to pursue include adjusting the tree-walk strategy when only a few particles need forces, accumulating forces on remote processors instead of using the software cache, and reorganizing the tree structure to avoid full tree builds on the smallest time steps. On the pure performance side, there is opportunity to take better advantage of both the on-chip vector units (e.g., SSE and AVX) and GPUs. There is also work to be done in taking full advantage of SMP nodes. Finally, there is a plethora of subgrid physics features in cosmological simulations, e.g., star formation, supernovae and black holes, that can be added to CHANGA. The close relation between GASOLINE and CHANGA will makes the addition of these features straightforward.

Acknowledgments

A number of people besides the authors contributed to the development of CHANGA. Graeme Lufkin wrote the initial version, and Sayantan Chakravorty wrote the original version of the software cache. Amit Sharma also contributed to domain decomposition and the gravity tree-walk. The hexadecapole moment calculations, including the Ewald expansions, were taken from PKD-GRAV and were mainly authored by Joachim Stadel. The SPH routines are based on the GASOLINE implementation written by James Wadsley. Lukasz Wesolowski contributed to the CUDA port. We also thank Orion Lawlor for many instructive discussions. Victor Debattista is also thanked for being a steadfast early adopter. We thank the reviewer for helpful comments that improved this chapter. Initial development of CHANGA was supported by NSF ITR grant PHY-0205413. We also acknowledge support from the NASA AISR program.

Chapter 7

Remote Visualization of Cosmological Data Using Salsa

Orion Sky Lawlor

Department of Computer Science, University of Alaska Fairbanks

Thomas R. Quinn

Department of Astronomy, University of Washington

7.1 Introduction

As discussed in the previous chapter, the experimental cosmology application CHANGA simulates particle datasets consisting of tens of millions to billions of particles. Salsa, the visualization tool we describe in this chapter, is the simulated observatory intended to help extract useful information from the simulation results.

There are two key challenges in extracting scientific results from simulations of cosmology. First, the natural dynamic range of cosmology runs from hundreds of megaparsecs, which are required to get a "fair sample" of galactic environments, to star formation environments in individual galaxies with scales of parsecs. This later scale is required because it is primarily the light from the stars or the supernovae by which we get the observational constraints on our cosmological models. Simulating such a large dynamic range leads to the generation of large datasets. The second challenge is that galaxies and galaxy clusters are non-trivial 3-dimensional objects. They sit in triaxial halos; they are characterized by their morphology which is a complicated interaction of components with different shapes; they are formed via a complicated merger process. Interpreting the flatness of the disk, the strength of the spiral arms, the size of the bulge, the presence of a bar and the warping and bending of any of these components requires the ability to grasp the full 3-dimensional structure of the galaxy. This information is most effectively conveyed to the investigator by rendering the model at a smooth interactive rate.

Unfortunately, these two challenges are incompatible at face value. The large dynamic range results in datasets that run into the terabytes, while interactive rendering is limited to a few gigabyte datasets even with high-end graphics workstations. The memory requirements alone require parallel architectures which do not lend themselves to interactive use. Furthermore these intrinsic problems are compounded by the fact that, in a typical situation, the simulation datasets are computed at a supercomputing center, far away from the application scientist and connected to him via a relatively slow internet link, running at only megabytes per second.

In this chapter, we show how we designed Salsa to visualize these large multi-gigabyte multidimensional datasets over a relatively slow network link, and provide interactive analysis capabilities to the scientist. This design makes use of a number of capabilities of the Charm++ framework, as well as the construction of a client application that exploits the graphics capabilities of desktop workstations. To demonstrate the utility of our work, we end with a description of how Salsa was used in a recent galaxy cluster cosmology project.

A separate data analysis challenge is to be able to interactively query this large simulation result for quantitative results. Furthermore, coupling such queries to the interactive visualization can be very productive; e.g, "What is the mass of the cluster that I just selected with the graphical interface?" To address this challenge, Salsa has the capability of both querying and manipulating the simulation data via a Python based interface. For a more complete description of this capability, we refer the reader to a previous publication [79].

7.2   Salsa Client/Server Rendering Architecture

Generally, ChaNGa is run on thousands of processors, which are needed to provide the computational capacity to perform the simulation in a reasonable time. Large processor allocations like this are typically batch scheduled, which is not convenient for interactive scientific data analysis. Luckily, data visualization has lower computational complexity than simulation, so we can normally load the simulation data onto fewer processors for visualization and analysis, typically a few dozen processors, small enough to be allocated for interactive use. For the analysis of a very large dataset, for example with hundreds of billions of particles, in principle we could load data across tens of thousands of processors, since our visualization infrastructure is designed

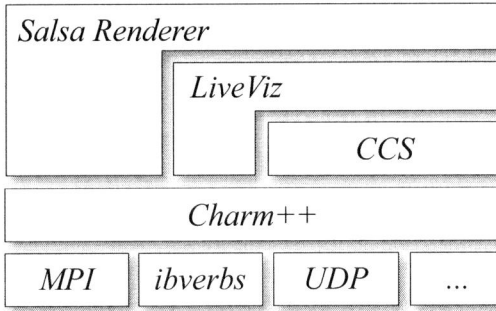

FIGURE 7.1: Software layer diagram for Salsa server-side renderer.

to scale well. But in any case, to display anything useful to the application scientist, we need to send data to a local client machine.

As shown in Figure 7.1, Charm++ provides several useful tools to solve this remote visualization problem. The Charm++ network protocol Converse Client Server (CCS) [113, 47] allows a Charm++ program to act as a network server and allow authenticated connections from clients via the network. On top of CCS, the application-layer image assembly protocol LiveViz [46] allows individual array elements to provide fragments of a 2D image, which LiveViz assembles across processors, compresses and transmits to the client via CCS.

Our overall data movement during runtime is thus shown in Figure 7.2. We typically need several dozen gigabytes of RAM to store the entire particle dataset, so we leave this parallel particle list distributed across the visualization cluster. The exact distribution of particles across processors can be transparently adjusted at runtime using the Charm++ object migration and load balancing libraries. To display this dataset on the client, we begin by rendering each array elements' particles into either a 2D image or a 3D volume impostor—a low-resolution stand-in for the actual data, described in the next section. These rendered images are then composited hierarchically at the core, node and inter-node levels across the cluster using the LiveViz library. The assembled image is then compressed and sent to the client, as described in Section 7.2.2. The Java client application maintains the CCS connection to the server, decompresses the received image data, and renders it using the local graphics card via JOGL, a Java binding for OpenGL graphics calls.

7.2.1 Client Server Communication Styles

In theory, the data analysis server could send any arbitrary data to the scientist's desktop machine. For example, the server could simply send fully-computed numbers, such as the measured galactic density fall-off exponent. For this style, Salsa supports a Python-based programming interface for writing and executing new data queries at runtime. The primary drawback of

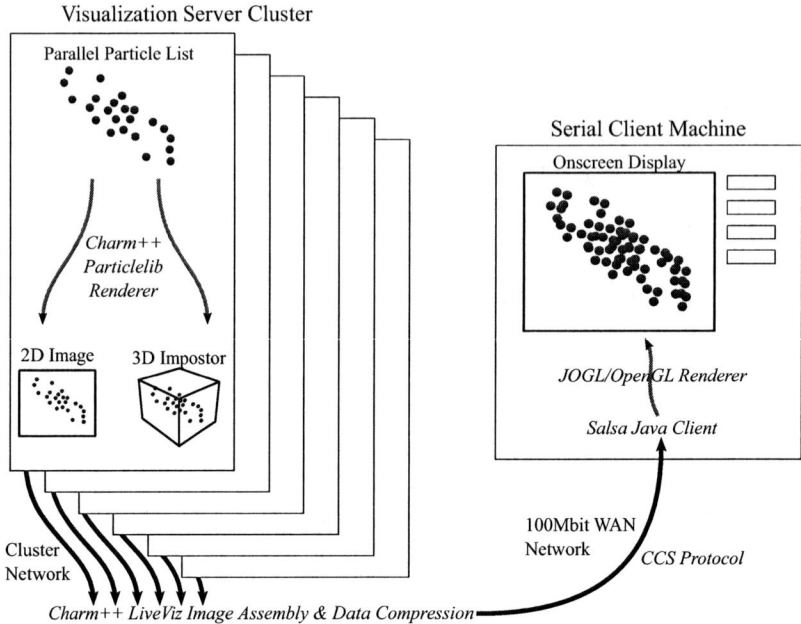

FIGURE 7.2: Data movement in Salsa. Particles are distributed across the cluster, rendered to 2D image or 3D impostor fragments, composited across the cluster, compressed and sent across the network to the client.

sending only numbers is summed up by Hamming's quote [97], "The purpose of computation is insight, not numbers." Visualization is useful because the story of scientific progress often begins with "Oh no!" rather than "Ah ha!" moments, and it isn't always obvious where to begin analysis. The questions are as important as the answers, when trying to discover aspects of a dataset that you don't yet know to look for.

The server could send simple 2D images of the particle dataset, as viewed from some particular 3D viewpoint and scale factor. This is currently our primary method of interacting with data in Salsa. Sending images is general purpose, and leverages the spectacular data analysis capacity available in the visual cortex of the human brain. But one difficulty with 2D images is that changing the 3D viewpoint requires a round-trip to the server—the client requests a new viewpoint, the server renders the dataset from the new viewpoint and the image must be compressed and transmitted back to the client. This process can take several seconds, which makes it difficult to understand the 3D structure of the data. Astrophysicists are used to this problem in observational data, where it is difficult to estimate the 3D structure of objects billions of light years away, but in a simulation we should be able to do better.

To allow the scientist to interactively move through a 3D dataset, the server could send a 3D representation of the data. For tiny datasets, we could simply download the entire dataset, but most realistic datasets require more storage and rendering capacity than are available on most desktop machines. Random subsampling is promising, but tends to over-represent dense regions near galaxy and supercluster cores, and under-represent more distant regions. Hybrid approaches, such as mixed volume/particle methods [158], create difficult to tune tradeoffs between the particle and volume areas.

We have extensively explored volume rendering, where the particle dataset is sifted into a 3D voxel grid, and the resulting volume dataset sent to the client, where it can be explored in 3D with no further communication. We call this approach "volume impostors" as a generalization of the well known 2D texture-based "impostors" technique [159]. There are several advantages to volume rendering: for example, volume datasets compress well, with a 128 million voxel 512^3 image compressing to only a few megabytes. Surprisingly, even a 2007 desktop graphics card can interactively render a dense 512^3 volume image at over 20 frames per second. Salsa currently supports 3D volume impostors, but there are difficulties with this approach. The graphics hardware can most efficiently render from a regular 3D grid, which spans only a limited region of space; this means to get more detail in one part of the simulation, another area must be clipped away. Salsa maintains a "point of interest" as the user pans and zooms, to reduce this effect. Further, when the 3D viewpoint is stationary, the client requests a perfect 2D rendering including all the particles, and displays this as soon as it arrives, combining the interactivity of the 3D volume with the quality of the 2D image (see Figure 7.3).

7.2.2 Image Compression in Salsa

For the long trip over the network to the client, LiveViz supports several image compression mechanisms. The CPU-limited[1] throughput and bandwidth savings for several possible image compression algorithms are summarized in Table 7.1. Overall, only run length encoding is efficient enough to be useful on a gigabit network (100MB/s), though on slower networks GZIP can be a net benefit. While lossy compression methods like JPEG can be competitive in time and space efficiency, the resulting distortions are usually not acceptable, especially when scientific data is plotted in false color. We have shown that a careful GPU implementation of run length encoding can double the compression throughput compared even to multicore [211], but normally rendering is the bottleneck.

---

[1]The CPU in this table is one processor of a 3.1GHz Intel Core i5-2400.

Major Viewpoint
Change (zoom)

New 3D
Impostor

3D Impostor
Complete

Minor Viewpoint
Change (rotate)

Reuse 3D
Impostor

Mouse
Released
(static view)

Client Idle

2D Image
Complete

New 2D
Image

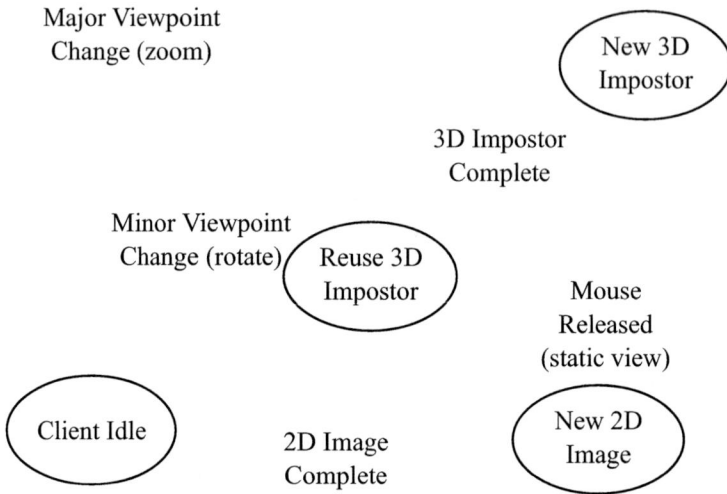

FIGURE 7.3: Finite state machine for network communication on the client.

7.2.3 GPU Particle Rendering on the Server

Since the server's job is to render a large array of particles into an image, it is natural to consider using the graphics processing unit (GPU) to accelerate rendering. There are a number of complications with this, however.

First, the particle data is loaded on the CPU, and must be copied to the GPU. This copy time exceeds the time to render the data, so this is affordable only if the time to copy the data onto the GPU can be amortized across several frames. Hence Salsa tracks changes to the particle data such as when the user selects a new attribute, coloring or dataset otherwise Salsa leaves the particle

Algorithm	Compress (MB/s)	Savings (%)	Decompress (MB/s)	SNR
Run Length (RLE)	91	40	125	*lossless*
DEFLATE (GZIP)	16	80	67	*lossless*
DEFLATE (PNG)	9	80	59	*lossless*
Discrete Cosine (JPEG)	25	75	33	20dB

TABLE 7.1: Compression algorithm comparison for Salsa server-to-client images. Performance is shown for a representative 1000×1000 pixel grayscale image.

GPU Data Writes	Performance (Gop/s)
Naive byte writes	3.1
Software atomic byte max	0.6
Naive int writes	4.9
Atomic int max	4.4

TABLE 7.2: Memory write performance on NVIDIA GeForce GTX 580 graphics card. Speeds are in billions of memory operations/second. Note that naive writes will produce incorrect answers when particles overlap.

data on the GPU. In particular, multiple 2D viewpoints or 3D zoom factors can be rendered by the GPU without needing to copy the data each time.

Second, because 3D particles can be drawn to arbitrary locations in the 2D framebuffer, it is not obvious how to render multiple points in parallel—there is a race condition whenever two points try to modify the same pixel. Although modern GPU hardware supports atomic memory operations quite efficiently, atomic operations are only available for integer or float data types. Typically images use byte pixels, for which the hardware does not provide atomic memory operations. We resolve this by promoting pixels to full integer width, which allows us to use the hardware native atomic operations. As shown in Table 7.2, this is a higher performance solution than simulating atomic byte memory operations in software.

Finally, write contention can be a significant GPU performance problem when several 3D points project to the same 2D pixel, requiring the parallel hardware to serialize the pixel writes at runtime. Astrophysics data, where points gravitationally cluster and collapse into very small regions, causes especially poor write contention when rendering the dense core of galaxies. On a single core CPU, this is not a performance issue, because the pixels covered by the galaxy core stay in cache; but these areas are a serious performance issue on either multicore CPU or the GPU.

We have found that declustering the original particle data, by randomizing with a Fisher-Yates shuffle during upload, reduces memory contention during the atomic pixel write operations. This, in turn, improves runtime rendering performance over threefold for modern GPU hardware. Surprisingly, this effect persists even in modern cached GPU memory architectures such as NVIDIA's Fermi. It is surprising that on modern parallel hardware, high data access locality can actually create contention, decreasing performance. Instead, randomizing memory accesses can reduce memory contention and improve performance—this is the exact opposite of the principle of locality to improve sequential cache performance!

7.3   Remote Visualization User Interface

Figure 7.4 shows the Salsa user interface. On the right is a Java Swing interface for selecting particle families and fields of the particle to visualize. On the left is a 3D rendering shipped across the network from the server.

An interesting network design issue arises when the user manipulates the 3D view. As the user drags the mouse to rotate the image, the operating system reports the mouse position about a hundred times per second. Although we can easily send out a hundred new image requests per second across the network, with billions of particles even a large parallel machine cannot render them this quickly, and often the network cannot deliver frames this rapidly either. This asymmetry makes it easy for the client to unintentionally overload the server and saturate its own network link; so, to maintain responsiveness, we must limit the number of image requests sent. A hardcoded or user-selectable limit on the number of outgoing requests per second does not scale well to the variety of parallel rendering hardware, particle counts and network links encountered in reality; but we find that a limit on the number of requests outstanding self-adapts to these limitations. With a limit of one outstanding request, we effectively have a "last in, only out" queue of size one. This ensures that as soon as the server is free, we immediately send it the most up-to-date request [146].

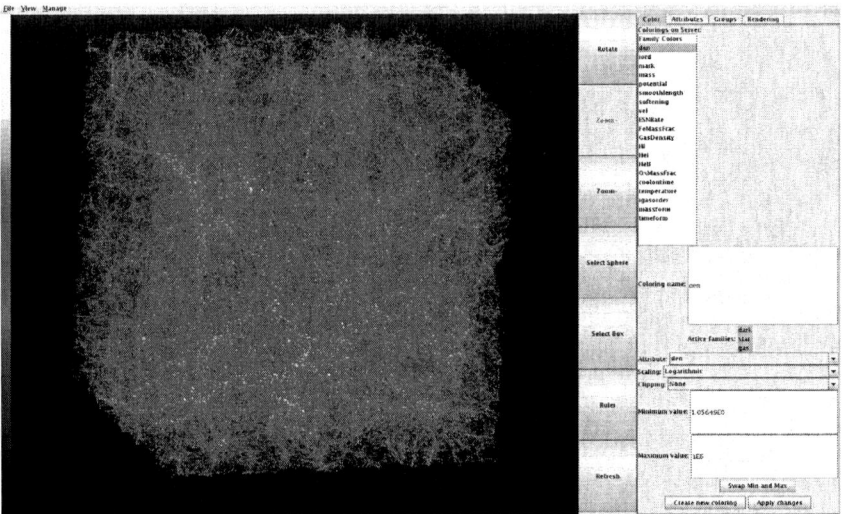

FIGURE 7.4: Screen user interface screenshot, showing the density field for a 2-billion particle simulation. The data is in Oak Ridge, Tennessee; the client is in Seattle, Washington.

7.4   Example Use: Galaxy Clusters

Here we give an example of the use of Salsa for a specific cosmology project: understanding observations of galaxy clusters. Clusters of galaxies are excellent probes of cosmology for a number of reasons. First, they are sensitive probes of the growth of cosmic structure because their number has evolved rapidly in the recent past. Second, since they are the largest gravitationally bound objects in the Universe, they can be observed over a large range of cosmic time. One effective method for detecting clusters at very large distances is through the distortions they produce in the Cosmic Microwave Background (referred to as the Sunyaev-Zeldovich Effect, SZE, [249]); however, many clusters found this way seemed to have disturbed morphology [198], perhaps indicating a selection bias. In the work described below, we use CHaNGa simulations of a galaxy cluster to follow the evolution of the SZE through the growth of a large galaxy cluster. This work is described more fully in [210]. Here we point out the use of Salsa at various stages in this project.

The first stage of simulating the growth of a galaxy cluster is picking a representative cluster. This usually requires simulating a "fair sample" of the Universe, identifying the clusters, and selecting one of the desired mass and formation history. Figure 7.5 shows a slice through such a simulation rendered with Salsa. Cluster identification was done by a separate program that implements the "friends-of-friends" algorithm [43] and produces a cluster ID for each particle. This program was run for each of 30 snapshots in time. Salsa ran a python script that successively read in each of these data files, calculated the cluster masses and, for each particle, calculated the ratio of the mass of the cluster it was currently in to the mass of the cluster it was in at the end of the simulation. A formation time was defined to be when that ratio crossed a chosen threshold (e.g., 75%). While this calculation did not need the visualization capabilities of Salsa, having a parallel framework controlled by a high level scripting language made it easy to program and quick to run.

Once a cluster of suitable mass and formation history is selected, the gas dynamics is followed in high resolution by determining the Lagrangian region from which the cluster formed and resimulating that region with much greater resolution. Salsa is helpful for several aspects of this process. First, determining the quality of the initial conditions involves verifying that the entire region from which the cluster formed is represented by high resolution particles, as well as a "buffer" region surrounding it. This is easiest done by visual inspection of the particle distribution. Then after the simulation is performed, its quality can again be assessed by investigating whether the cluster is "contaminated" by lower resolution particles. As well as visual inspection, quick subselections of particles via python scripts were used to determine whether this was the case. The final output of the gas dynamical simulation of the cluster is shown in Figure 7.6.

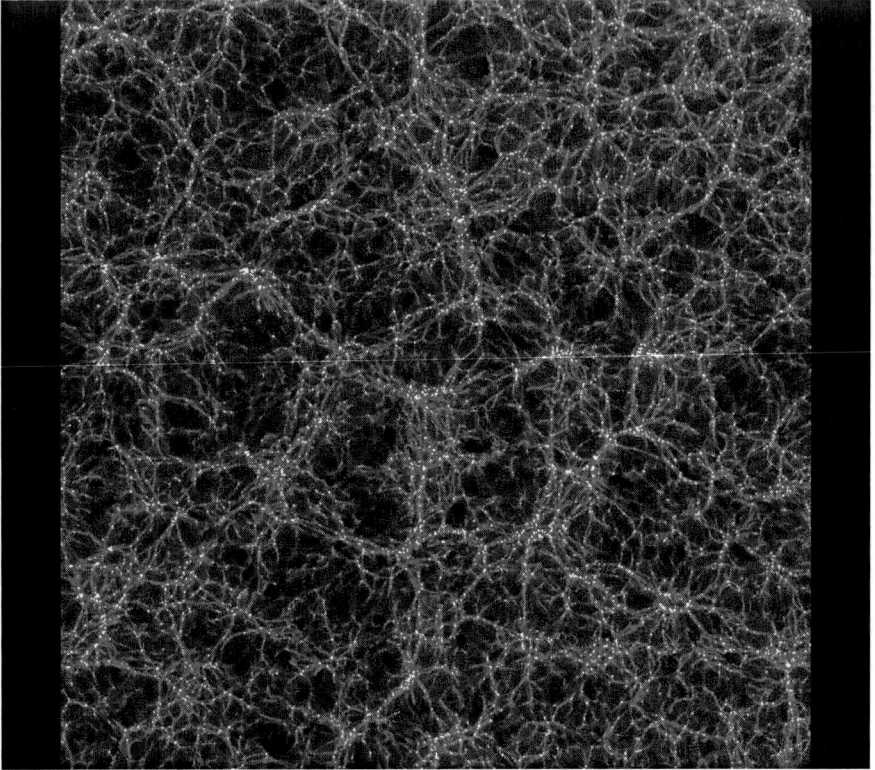

FIGURE 7.5: Salsa rendering of the dark matter density distribution in a simulation of galaxy cluster formation (**see Color Plate 6**).

Finally, the simulation is analyzed to produce the scientific results. The nature of the final merger that produced the cluster was determined by visually inspecting the cluster in a number of snapshots to identify the two pre-merger clusters, and calculating their mass. Some of the quantities to be measured are strongly dependent on the direction from which the cluster is viewed: substructure most easily identifiable when the apparent separation from the cluster center is maximized, a kinetic component of the SZE is a function of the line-of-sight velocity of the substructure. Salsa was used to interactively discover these optimum viewing angles. Simulated observations of the cluster were made with Python scripts that binned the particles into instrumental pixels and calculated the appropriate summation of the particle quantities.

FIGURE 7.6: Salsa rendering of the gas density within a cluster of galaxies (see **Color Plate 6**).

Acknowledgments

The construction of Salsa was supported by a grant from the NASA Applied Information Systems Research effort and a grant of HPC resources from the Arctic Region Supercomputing Center and the University of Alaska, Fairbanks.

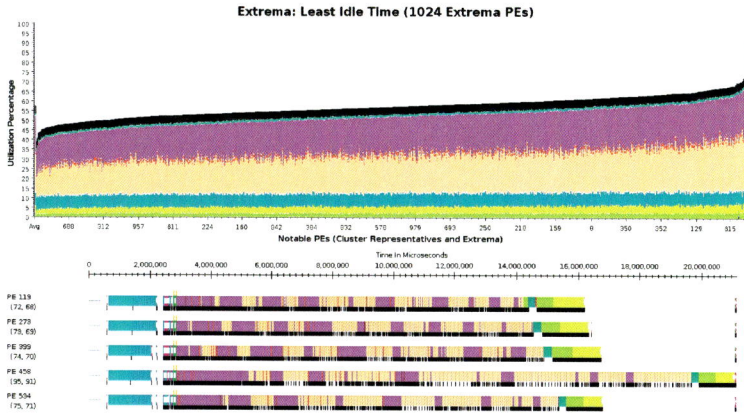

Color Plate 1. Extrema and timeline view of simulating a 5-million particles system in CHANGA using 1024 cores on Blue Gene/P (Figure 3.11).

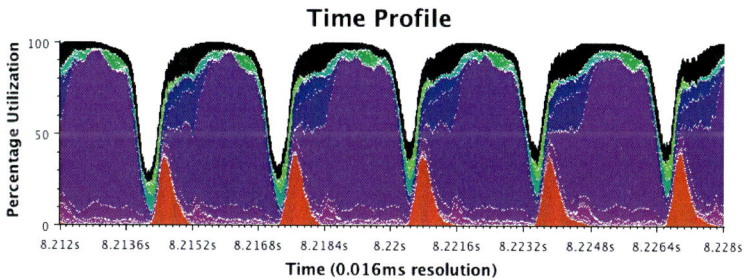

Color Plate 2. Time profile of simulating ApoA1 on 1024 cores with 2-away XY decomposition (Figure 3.14).

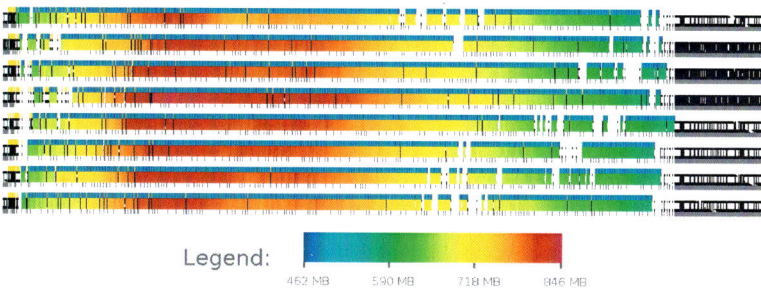

Color Plate 3. Memory usage as visualized by Projections via a modified timeline. Entry methods are colored according to their memory usage. A legend shows the color scale and the data values (Figure 3.16).

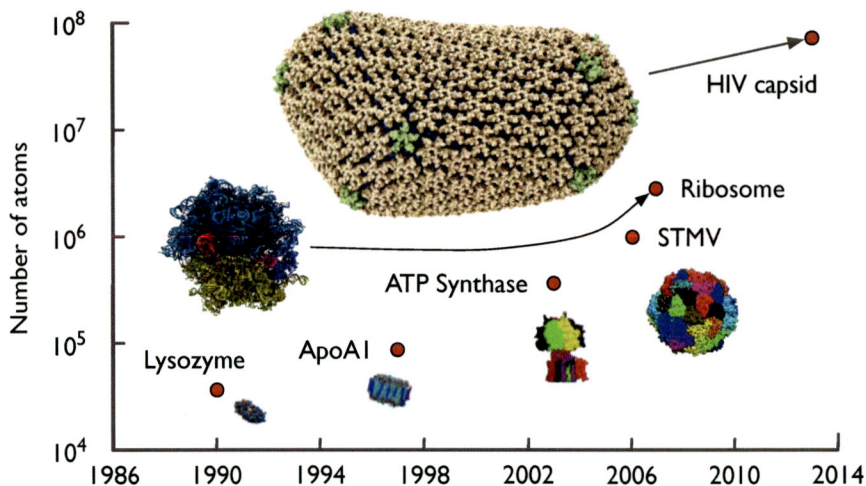

Color Plate 4. The size of biomolecular systems that can be studied using all-atom molecular dynamics simulations has steadily increased from that of a Lysozyme (40,000 atoms) in the 1990s to the F_1F_0-ATP synthase and STMV capsid at the turn of the century, and now 64 million atoms as in the HIV capsid model shown above. Atom counts include aqueous solvent, not shown (Figure 4.6).

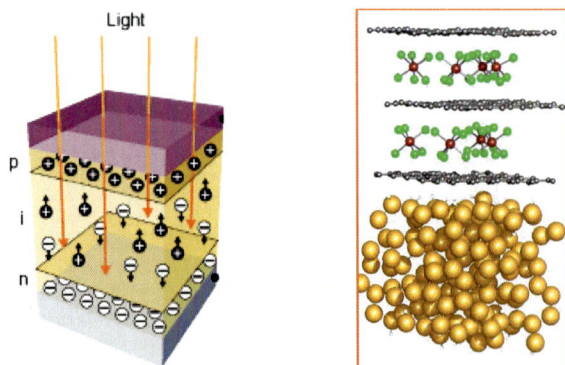

Color Plate 5. (Left) Mock-up of a solar cell with a transparent top electrode, a PIN junction and a bottom electrode. (Right) A snapshot of a simulation of (from bottom to top) amorphous silicon terminated by hydrogen atoms, graphene, a layer of antimony pentachloride (SbC 15), graphene, a layer of antimony pentachloride ($SbCl_5$), graphene (Figures 5.7 and 5.9).

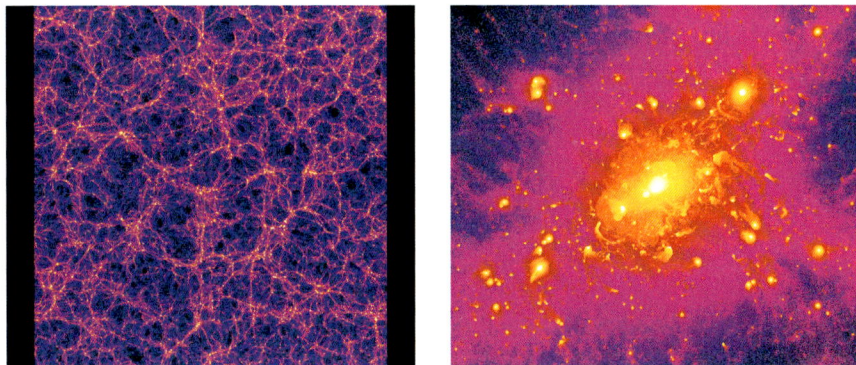

Color Plate 6. Salsa rendering of the dark matter density distribution in a simulation of galaxy cluster formation (left, Figure 7.5), and Salsa rendering of the gas density within a cluster of galaxies (right, Figure 7.6).

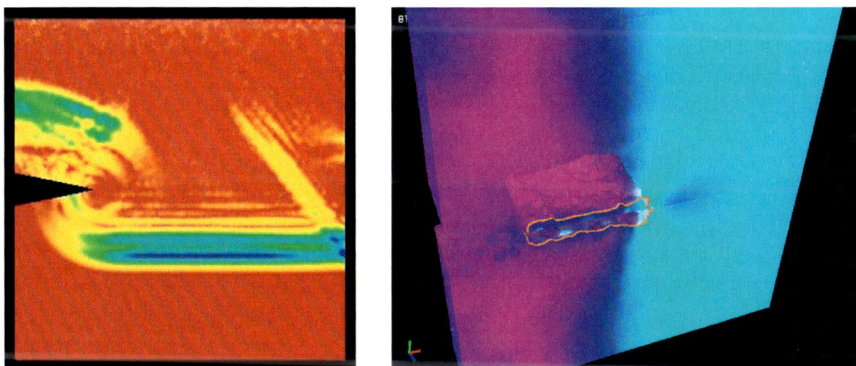

Color Plate 7. Diffraction off a crack, simulated in parallel and rendered using NetFEM directly from the running simulation (left, Figure 9.3); Snapshot of the plastic zone surrounding the propagating planar crack at time $c_d t/a_0 = 27$. The iso-surfaces denote the extent of the region where the elements have exceeded the yield stress of the material (right, Figure 9.13).

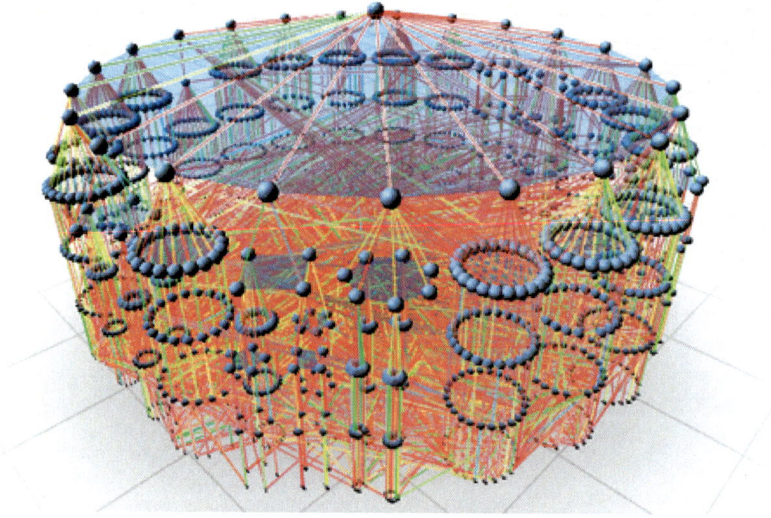

Color Plate 8. Example of a synthetic person-person contact network, where people are represented by balls, the size of which is inversely proportional to its degree of separation from the top person. Two people who come in contact during the day are linked by an edge (Figure 10.1). (Image courtesy of David Nadeau, SDSC.)

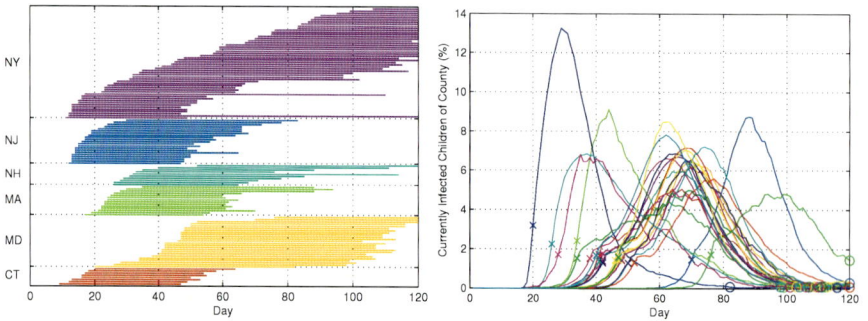

Color Plate 9. The school closure timing by county, sorted by initial closure day within each state (left, Figure 10.18); the percentage of infected children in each county of MD when school closure intervention is applied. The day when the school closure begins effective for each county is shown by ×, and reopening is shown by ○ (right, Figure 10.18).

Chapter 8

Improving Scalability of BRAMS: a Regional Weather Forecast Model

Eduardo R. Rodrigues

IBM Research Brazil

Celso L. Mendes

National Center for Supercomputing Applications, University of Illinois at Urbana-Champaign

Jairo Panetta

Instituto Tecnologico de Aeronautica – ITA

8.1 Introduction

Numerical weather forecasting models have a long history. Currently they have become important decision tools in many different areas, ranging from agriculture to flight control and energy production. Many applications require forecasts with very high resolution. To execute these models in a feasible amount of time, they are typically run in parallel.

Similarly to other applications, many issues can prevent weather models from fully exploiting all the capability of a parallel computer. One such issues is load imbalance, i.e., the uneven distribution of load across the processors of a parallel machine.

Load imbalance in weather models can be caused by either static or dynamic factors. An example of a static factor is topography. Many weather models represent the atmosphere with a three-dimensional grid of points, distributing those points across the processors according to a horizontal domain decomposition in the latitude/longitude plane. Each processor handles all the atmospheric columns corresponding to the horizontal points in its partition. The grid cells may not follow the terrain (for example, in a shaved cell representation of topology [1]) and, therefore, some cells may not be active. Therefore, locations with a high topography result in fewer active cells, hence less work to be computed. Such imbalance does not change for a given region; therefore, if the model is routinely used on the same region it is feasible to derive, in advance, an ideally balanced decomposition.

The dynamic factors leading to load imbalance in weather models are much more complex. Some of these dynamic factors are predictable, whereas others are unpredictable. One example of a predictable factor is radiation. Two distinct regions are subject to different values of solar incidence at a given time (e.g., day and night); this results in distinct amounts of computation for the radiative components of the model on the two underlying processors. This kind of imbalance repeats with a periodicity of twenty-four hours in simulated time. To deal with this factor, Foster and Toonen devised a strategy that redecomposed the domains, keeping the amount of work across processors even throughout that period [66].

However, other imbalance factors may not present the same predictability. As an example, running the weather model described in Section 8.3 on 64 processors produced the precipitation forecast shown in Figure 8.1(a). By instrumenting that execution, we captured the computational loads in the 8×8 grid of processors, as depicted by the grayscale-coded representation of Figure 8.1(b). The darker regions in Figure 8.1(b) correspond to higher computational loads. One can clearly see a strong correlation between rain and computational load, as regions with rain correspond to processors that are overloaded.

Running the simulation of Figure 8.1(a) for a longer period of time results

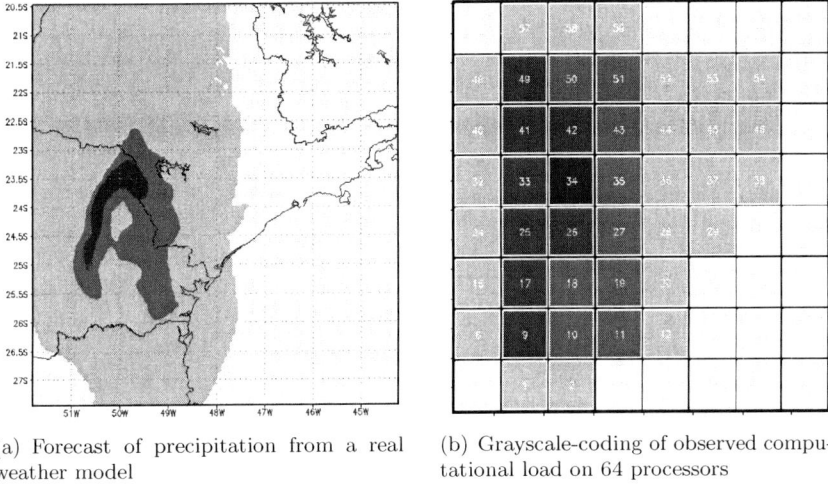

(a) Forecast of precipitation from a real weather model

(b) Grayscale-coding of observed computational load on 64 processors

FIGURE 8.1: Example of dynamic load imbalance.

in a "movement" of the rain across different regions, which in turn changes load distribution over time. The precise path that the rain follows is unknown a priori (since predicting where there will be rain is what the model is designed for!). Handling these unpredictable sources of load imbalance in weather models appropriately (i.e., maintaining accuracy in the forecast) and efficiently (i.e., without increasing the computational cost of the model) remains mostly an open problem.

In this chapter, we will describe how Adaptive MPI (AMPI) was used to deal with the load imbalance of a real regional weather model—BRAMS (Brazilian developments on the Regional Atmospheric Modeling System) [27]. In addition, the over-decomposition strategy in which AMPI is based helped to improve communication performance. Here, we also show results of this extra benefit. The remainder of this text is organized as follows. The next section presents a review of previous work in the area. Section 8.3 describes the weather model we used in our study. It is followed by our approach for doing load balancing in weather models. This approach is based on processor virtualization and its implementation on AMPI. We also describe our load balancer, which is based on the Hilbert curve. Finally, we present experimental results and concluding remarks.

8.2   Load Balancing Strategies for Weather Models

Embedding the balancing strategy into the application itself has been the most common way to implement load balancers [171] [77] [66]. Many examples of such approach can be found in the literature. For instance, the Gordon Bell winner of 2002 [220] embeds a load balancing strategy in a spectral atmospheric general circulation model (Atmospheric Model for the Earth Simulator, AFES). Its approach is to rebalance load using a particular characteristic of this method. This strategy is not general; it can only be applied to spectral models.

Koziar et al. [136] study load imbalance of a regional weather model named Gesima. That article notes that, although weather models have a regular structure, atmospheric processes can cause load imbalance throughout the domain. The objective of that study is to select criteria to activate microphysics so that the results are correct, but also to balance load across the processors. However, that work considers only one source of load imbalance; it does not deal with the composition of effects (for example, microphysics and the remainder of the model). In addition, that article does not perform actual load balancing; it only evaluates possible directions to adapt the application to deal with this problem.

MM90 is an example of a meteorological model that contains dynamic load balancing. It is a Fortran90 parallel implementation of the Penn State/NCAR Mesoscale model (MM5). Michalakes [171] presents how the MM90 was parallelized, a dynamic load balancing strategy and performance results. The domain decomposition is done in two dimensions (north/south and east/west). The sub-domains can be irregular and the processing unit is a mesh point. This approach makes load balancing easier, since any mesh point can migrate from one processor to another in order to rebalance load, even if the sub-domains become irregular. The MM90 code is instrumented so that the load balancing strategy makes an estimate of the imbalance. This instrumentation basically measures the computational cost of the vertical atmospheric columns. Periodically, a new mapping of the domain is computed and its efficiency is compared to the previous mapping. The article, however, does not describe how the new mapping is done nor how the performance results are compared. According to Rotithor [209], these two issues are critical to assess the efficiency of the load balancing strategy.

The spatial resolution of meteorological models is limited by the computing power available (with the exception of purely meteorological factors such as inadequate observation data). In order to avoid this restriction, the user can use downscale techniques. Ghan et al. [76] present a new downscale technique that uses orography to improve the resolution of the NCAR/CAM3 model (National Center for Atmospheric Research/Community Atmosphere Model). This scheme improves resolution but causes load imbalance.

Ghan and Shippert [77] present a load balancing algorithm for the downscaling technique based on orography. That article shows that a static load balancer can be used, because the elevation classes do not change. The proposed algorithm considers not only load but also communication. Since this is a static load balancer, the strategy cannot be used in a dynamic context. Furthermore, this algorithm is specific to the downscaling technique used.

Foster and Toonen [66] identify that physics computation is a source of load imbalance in climate codes. Examples of physics computation are radiation, which depends on the movement of the planet, and cloud and moisture, which are transported with the movement of the atmosphere. They propose a dynamic load balancing scheme based on a carefully planned exchange of data across processors at each time step. The rationale for this approach is that the model employs three types of time steps, with varying degrees of radiative calculations, and a good decomposition for one kind of time step is not as good for the other kinds. They achieved an overall improvement of 10% on 128 processors with the PCCM2 climate model which degraded with more processors. This technique has a significant execution overhead, as it requires a large amount of data exchange between processors at each time step. As the model is scaled to a growing number of processors, this overhead may dominate execution and offset any potential gains provided by the load balancing scheme. Furthermore, to implement this scheme the programmer needs intimate knowledge of the application's code, to determine precisely which variables must be exchanged between processors.

Xue et al. [247] indicate that active thunderstorms may cause the processors assigned to certain sub-domains to incur 20%-30% additional computation. They also claim that the complexity of the associated algorithm and the overhead imposed by the redistribution of load prevent the use of load balancing techniques.

8.3 The BRAMS Weather Model

Differently from previous work (some of which is described in the previous section), we do not embed the load balancer in the application. To test this approach, we chose a weather forecasting model for our experiments. In this section, we describe the weather model we used in our investigation. BRAMS (Brazilian developments on the Regional Atmospheric Modeling System, RAMS) is a multi-purpose regional numerical prediction model. It was designed to model atmospheric circulations at many scales. Currently, it is employed around the world for both production and research. BRAMS has its roots on RAMS [241], a model that solves the fully compressible non-hydrostatic equations described by Tripoli and Cotton [237]. It employs a multiple grid scheme with nested meshes of increasing spatial resolution. This scheme allows the model equations to be solved simultaneously on any number of two-way interacting computational meshes. There is a set of state-of-the-art physical parameterizations that are appropriate for the simulation of important physical processes such as surface-air exchanges, turbulence, convection, radiation and cloud microphysics.

Although BRAMS started as a research project to tailor RAMS to the tropics, it also aimed at modernizing the original software structure in RAMS.

Its modeling features extended RAMS in the following ways: (a) inclusion of a cumulus convection representation, as part of an ensemble version of deep and shallow cumulus scheme based on the mass flux approach [90]; (b) daily soil moisture initialization data [75]; (c) a specific surface scheme that allows the representation of important tropical phenomena; and (d) a coupled aerosol and tracer transport model (CATT-BRAMS [68]) that was developed for the study of emission, transport and deposition of gases and aerosols associated with biomass burning, such as those originated in the Amazon. CATT-BRAMS has been used in daily production mode at CPTEC to forecast air quality for the entire South America (see http://meioambiente.cptec.inpe.br/).

BRAMS employs Fortran90 features to eliminate dusty deck software constructs from the original RAMS code, such as static memory allocation and extensive use of Fortran77 commons. This enabled BRAMS to achieve production code of good quality while maintaining flexibility to keep it useful for research. The BRAMS code is open source and it is freely available at http://brams.cptec.inpe.br/. It is supported and maintained by a modest software team at CPTEC that continuously transforms research contributions into production quality code to be incorporated at future code versions. BRAMS also became a good platform for computer science research in themes such as grid computing [58, 225].

This work uses the current research version of BRAMS (v. 5.0). This version has enhanced parallelism when compared to the current production version. Up to the current production version, BRAMS still used the original master-slave parallelism from RAMS, where the horizontal projection of the 3D domain was partitioned into rectangles as close to squares as possible, assigning each rectangle to one slave process. The current research version removes the master-slave parallelism to avoid memory contention on the master process. The resulting code eliminated the memory bottleneck in the master, while enhancing parallel scalability up to a few thousand processors [62]. Load balancing then became the major scalability bottleneck, partially due to the rectangular domain decomposition but mainly due to the dynamic load variation during integration.

8.4   Load Balancing Approach

This section describes our methodology to balance load in the BRAMS meteorological model. That methodology relies on the concept of processor virtualization and its implementation on AMPI. We present the changes that are needed for a real-world application to use the migration capabilities of AMPI. However, simply enabling an application to run on AMPI is not enough to guarantee good load balance, as a load migration strategy must be chosen to fit the needs of a specific application. We started our tests with existing AMPI load balancers that could in principle be useful for meteorological mod-

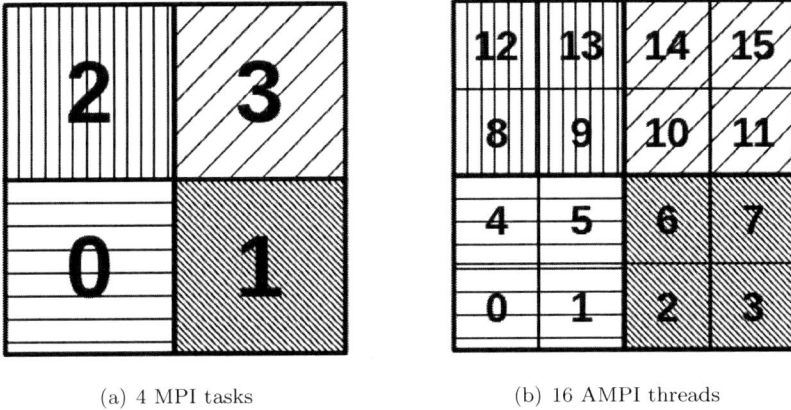

(a) 4 MPI tasks   (b) 16 AMPI threads

FIGURE 8.2: Domain decomposition.

els. The obtained results, which we present in Section 8.7, showed that such balancers did not produce a balance with sufficient quality. For that reason, we developed a new load balancer based on a heuristic that implicitly considers the communication pattern typically found in meteorological models. Furthermore, the algorithm is cheap enough to be invoked frequently. Initially, we implemented this load balancer in a centralized form, i.e., a single processor makes load balancing decisions. This approach works well for moderately large machines. For larger machines, we implemented a distributed load balancer based on the same heuristic, as presented at later sections of this chapter.

8.4.1   Adaptations to AMPI

To benefit from the advantages of AMPI, an existing MPI code must exploit processor virtualization. That can be done by replacing the usual task decomposition of MPI with a new scheme that over-decomposes the same domain into a larger number of sub-domains, which are then implemented with AMPI threads. Each AMPI thread runs as if it had its own processor, and that is the reason why the AMPI thread is said to run in a Virtual Processor (VP) in the Charm++ terminology. This process is illustrated by Figure 8.2: the left part of the figure shows a regular domain decomposition with four MPI tasks and four processors (the different hatched areas), whereas the part on the right of the figure shows a possible decomposition of the same domain into sixteen sub-domains corresponding to AMPI threads. Thus, under AMPI there are sixteen ranks. The virtualization ratio is four, i.e., there are four VPs in each real processor. Different values of VP and corresponding virtualization ratios can be used as well. This AMPI execution looks like an MPI execution with sixteen ranks, except that it is done on four physical processors.

To employ the over-decomposition scheme of AMPI, one must consider two factors: (a) the code must present binary reproducibility, meaning that the numerical results are independent of the number of MPI ranks used; and (b) there may be an increase in memory footprint due to over-decomposition, as both ghost zones and stack regions increase when more sub-domains are used. Nevertheless, the performance gains from virtualization are frequently high enough to justify the use of over-decomposition.

Considering the example of Figure 8.2, the AMPI execution starts with the first processor hosting threads {0,1,4,5}, the second processor hosting threads {2,3,6,7}, etc. When the load balancer is invoked at some point of the execution, threads may migrate from a processor to another, depending on the observed load behavior prior to the invocation. The precise movement of threads is a function of both the observed loads from each thread and of the balancing policy in use. Because many meteorological codes follow an iterative execution scheme, the load balancer invocation can typically happen after some number of time steps in the execution. For the AMPI execution represented in Figure 8.2, a physical processor is shared by four threads. Those four threads are contained in the same process, hence they share the global and static variables of that process. This kind of variable sharing may lead to problems if those variables are not handled correctly. For platforms supporting the ELF (Executable and Linkable Format) format [153], AMPI provides a build-time flag (*-swapglobals*) that automatically privatizes global variables to each thread. This flag causes each thread to have a private instance of each global variable. During context-switch, the Charm++ runtime system automatically switches those instances, according to the thread that will execute next. Unfortunately, this method does not apply to static variables. A possible solution to handle static variables is to create a module and include all statics in this module, adjusting corresponding declarations of those variables in the source code. This scheme turns statics into globals, allowing *-swapglobals* to be used to handle those variables.

Despite allowing effective privatization of global variables, the *-swapglobals* scheme may lead to low efficiency in some codes. This is due to the fact that this scheme makes the context-switch time to become proportional to the number of global variables in the code. With *-swapglobals*, the code must be compiled as shared library, and the linker creates a global offset table (GOT) containing pointers to all global variables in the code. Privatization is implemented by the Charm++ runtime system, at context-switch, by changing every entry of the GOT with data from the thread being resumed. Hence, for codes with many variables of this type, the context switch time may become excessively large. Table 8.1 shows the numbers of global and static variables for BRAMS and WRF, two popular weather forecasting codes. For codes like these, an update to the GOT becomes a very expensive operation.

To avoid the potentially large context switch overhead imposed by *-swapglobals*, we developed a privatization scheme [204] based on Thread-Local Storage (TLS). Traditionally, TLS has been used as a privatization mechanism in kernel-level threads. With TLS, the specifier *__thread* found in the C/C++

Model	Globals	Statics
BRAMS	10237	519
WRF-v.3	8731	550

TABLE 8.1: Number of global and static variables in two weather models.

language can be used to mark variables that must be thread-private. To extend this mechanism to AMPI's user-level threads, we adapted the Charm++ runtime environment such that the marked variables are kept private to each thread. We also made changes to the *gfortran* compiler such that global and static variables in BRAMS could be handled by our TLS scheme. Results from the use of this technique are in Section 8.7.1.

8.4.2   Balancing Algorithms Employed

Given the existing set of load balancers available in Charm++, we analyzed the following balancers: *GreedyLB*, *RefineCommLB*, *RecBisectBfLB* and *MetisLB*. *GreedyLB*, as implied by the name, takes a greedy approach and assigns the thread with the heaviest computational load to the least loaded processor. This assignment is repeated until the set of threads is exhausted. Obviously, no communication information is considered; hence, two threads that communicate intensively may end up in distinct processors, which is clearly undesirable. However, in view of the simplicity of this policy, the balancing process is often very fast.

The *RefineCommLB* balancer considers both computational load and communication traffic. Like *GreedyLB*, it moves objects away from the most overloaded processors to reach average, but it also attempts to preserve locality of communication between threads. It also limits the number of migrations, regardless of the observed loads. This balancer is typically used in cases where imbalances are not too high, and moving just a few threads is sufficient to make the load distribution even.

RecBisectBfLB is a balancer that recursively partitions the communication graph of threads employing a breadth-first enumeration. The partitioning is based on the computational loads of the threads, and proceeds until the number of partitions is the same as the number of available processors. The communication traffic is considered in the partitioning; however, there is no guaranteed minimization of the communication volume across partitions.

Finally, *MetisLB* is a balancer based on Metis [129], a well-known graph partitioning method. It works by partitioning the thread communication graph, considering both the computational load and communication pattern observed in the execution.

All of these Charm++ balancers employ a centralized approach, i.e., all load balancing decisions are made in one processor and communicated to the other processors. This should work well for a moderate number of processors.

However, as we show in the result section, from all load balancers employed, only *MetisLB* performed well in our experiments. Meanwhile, the domain decomposition typically employed in meteorological models can be leveraged in a new load balancer.

The BRAMS weather model, like many other weather forecasting models, employs a spatial domain decomposition. This decomposition is 2D in the horizontal plane. Each rank receives the full atmospheric columns corresponding to the points in its domain. There is a high volume of communication between ranks from sub-domains that are neighbors. Hence, mapping two ranks from neighbor sub-domains (and their associated threads) to the same physical processor will ensure that their communication is local to that processor, which minimizes the communication overhead.

We present a new load balancer based on a space-filling curve in the next section. This load balancer is appropriate for the type of domain decomposition found in meteorological models, since it keeps neighbor threads close together. As shown in Section 8.7, its results make it competitive to MetisLB, achieving similar performance at a lower balancing cost.

8.5   New Load Balancer

To further investigate the load balancing problem in the specific case of a two-dimensional domain decomposition, such as in weather models, we developed a new Charm++ load balancer based on a space-filling curve. It places the various threads with their observed loads into a two-dimensional Hilbert space-filling curve [101], and then iteratively cuts that curve into segments of nearly equal load until the number of segments is equal to the number of processors. With this process, threads corresponding to sub-domains that are close in space are likely to be assigned to the same processor. Because typically there is a larger proportion of communication between neighbor threads, a significant amount of communication will become local to the processor. This, in turn, leads to lower communication cost and better application performance. Figure 8.3 shows the Hilbert curve for a 4×4 domain decomposition.

Firstly, the new load balancer has to compute the mapping between the 2D domain decomposition and the Hilbert sequence. Liu and Schrack [157] describe an efficient coding scheme that we used as a subroutine of our load balancer. Liu and Schrack's algorithm encodes the sub-domains as follows: for a domain decomposition of size $2^r \times 2^r$, the sub-domain at location (x, y), $((x_{r-1}...x_1x_0)_2, (y_{r-1}...y_1y_0)_2)$ in binary, corresponds to the Hilbert sequence that is represented by a quaternary digit string $h = (q_{r-1}...q_1q_0)_4 = \sum_0^{r-1} 4^i q_i$ where $q_i \in \{0, 1, 2, 3\}$. Each quaternary digit h_k in h is represented by two bits h_{2k+1} and h_{2k}. These two bits are computed by the following recursive

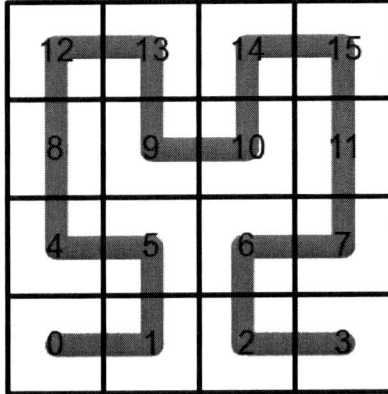

FIGURE 8.3: Hilbert curve for the case of 16 threads.

formulas:

$$h_{2k+1} = \bar{v}_{0,k}(v_{1,k} \oplus x_k) + v_{0,k}(v_{1,k} \oplus \bar{y}_k)$$
$$h_{2k} = x_k \oplus y_k$$

where $k = 0, 1, ..., r-1$ and the values of $v_{0,k}$ and $v_{1,k}$ can be computed by:

$$v_{0,r-1} = 0$$
$$v_{1,r-1} = 0$$
$$v_{0,j-1} = v_{0,j}(v_{1,j} \oplus \bar{x}_j) + \bar{v}_{0,j}(v_{1,j} \oplus \bar{y}_j)$$
$$v_{1,j-1} = v_{1,j}(x_j \oplus y_j) + (\bar{x}_j \oplus y_j)(v_{0,j} \oplus \bar{y}_j)$$

where $j = r-1, ..., 2, 1$.

To decode the Hilbert sequence h back to a sub-domain (x, y) the following formulas are used:

$$x_k = (v_{0,k}\bar{h}_{2k}) \oplus v_{1,k} \oplus h_{2k+1}$$
$$y_k = (v_{v0,k} + h_{2k}) \oplus v_{1,k} \oplus h_{2k+1}$$

where $v_{0,k}$ and $v_{1,k}$ are computed by:

$$v_{0,r-1} = 0$$
$$v_{1,r-1} = 0$$
$$v_{0,j-1} = v_{0,j} \oplus h_{2j} \oplus \bar{h}_{2j+1}$$
$$v_{1,j-1} = v_{1,j} \oplus (\bar{h}_{2j}\bar{h}_{2j+1})$$

According to Liu and Schrack [157] this algorithm is $\mathcal{O}(r)$. Since the dimension of the domain is $2^r \times 2^r$, the complexity with respect to the number of threads is $\mathcal{O}(\log N)$.

After computing the Hilbert sequence, the load balancer labels the set of threads with their loads. Afterwards, the sequence is cut in *numPEs* (total number of processor) segments with approximately the same load. This routine is described in Algorithm 1. The algorithm receives as input the observed load (*load[N]*) of each one of the N threads and the total number of processors (*numPEs*). It computes the prefix sum of the load, which is used to find the best cut of the Hilbert sequence, i.e., a cut that is closest to the ideal load balancing. The output is the vector *cutVec* that marks the places on the Hilbert sequence that delimit each segment. These segments are assigned sequentially to the available processors. Figure 8.4 shows an example of a possible assignment. As it can be seen, the threads in each segment are close together; therefore, the external communication is reduced.

> **input** : **float** load[N]
>   **int** numPEs
> **output**: **int** cutVec[N]
>
> prefixSum[0] = load[0];
> **for** *(i = 1; i < N; i++)* **do**
> | prefixSum[i] = prefixSum[i-1] + load[i];
> | cutVec[i] = False;
> **end**
>
> idealLoad = prefixSum[N-1] / numPEs;
> **for** *(i = 1, j = 0; i < numPEs; i++)* **do**
> | *inner*: **for** *(; j < N; j++)* **do**
> | | **if** *(idealLoad * i - prefixSum[j+1])²* > *(idealLoad * i - prefixSum[j])²* **then**
> | | | cutVec[j + 1] = True;
> | | | **break** *inner*;
> | | **end**
> | **end**
> **end**

Algorithm 1: Cut algorithm.

In spite of its simplicity, the new load balancer based on the Hilbert curve was very effective, as demonstrated by the performance results shown later in this chapter. However, the original algorithm employed by this balancer has a strong restriction related to domain geometry: the shape of the domain must be a square and its side must be a power of two. Chung, Huang and Liu [38] proposed an algorithm to overcome this limitation, allowing the use of rectangular domains with any size. While those authors presented that algorithm for image processing applications, we used the same algorithm to handle the load balancing problem.

First, the algorithm finds the largest square inside the original domain and places this square at the upper left corner of the domain. This step is applied recursively to the remaining area of the original rectangle, as illustrated in Figure 8.5(a). Each remaining region of the previous step is further decom-

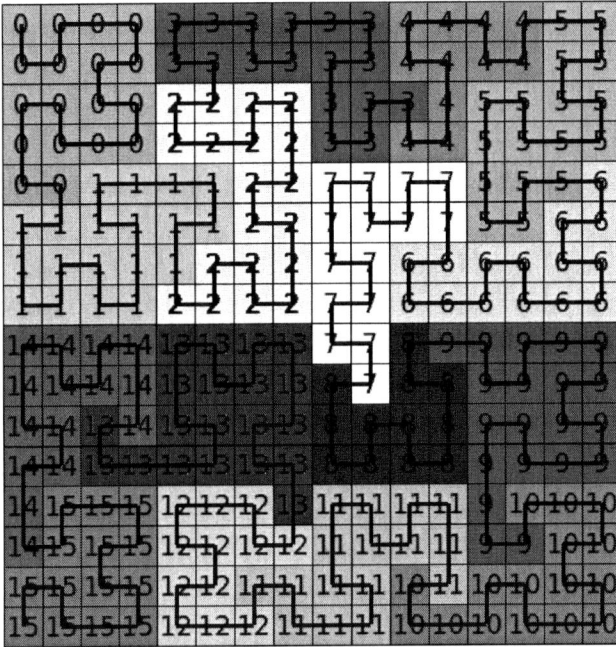

FIGURE 8.4: 16 processes and 256 threads after rebalancing.

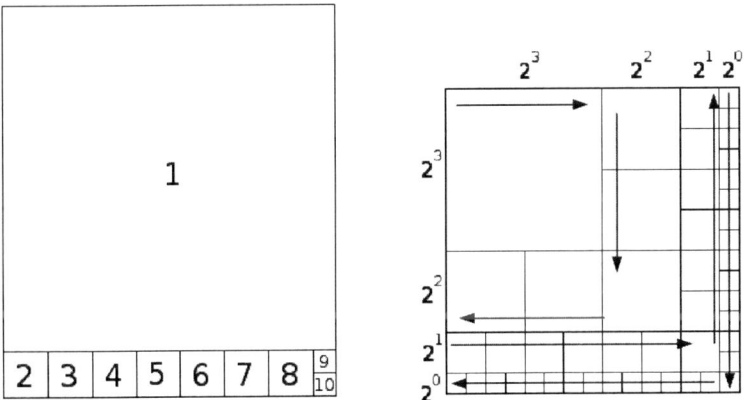

(a) Divide the original domain into squares

(b) For each square, apply the "snake-scan" approach

FIGURE 8.5: Decomposition of domains of arbitrary size.

FIGURE 8.6: Hilbert curve for domains of arbitrary size.

posed into smaller squares whose side is a power of two. To achieve that, a "snake-scan" approach is used, as shown in Figure 8.5(b). Finally, each smaller square is traversed by the regular Hilbert curve following the direction used in the previous step (Figure 8.6).

8.6   Fully Distributed Strategies

In the previous section, we discussed centralized load balancers. This type of load balancer has some advantages. Firstly, they are easy to implement, since all the information about load is gathered at one processor. Secondly, the balancing quality depends exclusively on the algorithm itself, because it has a complete view of the load distribution. Nonetheless, this approach has a major issue: it does not scale. Of course, an alternative is a distributed load balancer.

This section describes two fully distributed load balancers that we developed and evaluated. The first one is an extension of the load balancer that we presented in Section 8.5. The second strategy is based on the principle of diffusion.

8.6.1   Hilbert Curve-Based Load Balancer

A central entity that receives all load information and distributes migration decisions is a bottleneck of the strategy described in the previous sections. For

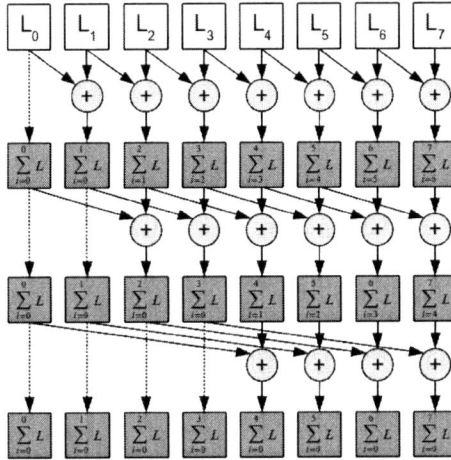

FIGURE 8.7: Parallel prefix sum.

moderately large machines, this is not a problem, because the load balancer may not be called very frequently. However, with many thousands of threads and a higher invocation frequency, the load balancer execution may cause a very large overhead. Fortunately, the Hilbert-curve based load balancer can be fully distributed and, therefore, this bottleneck is eliminated.

The first step of the distributed load balancer is to compute the Hilbert sequence. Liu and Schrack's algorithm (described in Section 8.5) can be used, since each thread can encode and decode the Hilbert sequence independently. All threads need to use this algorithm only once, because this process is necessary only at the beginning of the execution. Moreover, each thread does not need to compute the whole sequence, but only its position on the sequence and the position of those threads that communicates with it.

The next step is to compute the prefix sum of the loads of all threads. The prefix sum is an operation that takes as input a list and produces a result list in which each element is obtained from the sum of the elements in the operand list up to its index. The recursive doubling algorithm can be used to perform this list operation in $log_2(N)$ steps (where N is the number of threads) [107]. This algorithm is illustrated in Figure 8.7. At the end of its execution, each thread will have its corresponding element of the result list.

The third step is a broadcast. The last thread (thread $N-1$) has to send the total load to all the others; this thread has this information as a result of the previous step. This operation can also be performed in $log_2(N)$ operations. This step is needed so that each thread can compute the ideal load, which is given by the total load divided by the number of processors.

The final step is to execute a routine corresponding to Algorithm 2. This routine can be performed by each thread independently, because a thread needs only its own prefix sum element, its load, the total load and the number of processors. The result is the processor (*DestPE*) to where the thread must migrate.

input : **float** myPrefixSum
  float myLoad
  float totalLoad
  int numPEs
output: **int** destPE

idealLoad = totalLoad / numPEs ;
destPE = ⌊ myPrefixSum / idealLoad ⌋ ;
destPELeftNeighbor = ⌊ (myPrefixSum - myLoad) / idealLoad ⌋ ;
if *destPE ≠ destPELeftNeighbor* **then**
  if *(idealLoad * destPE - myPrefixSum)²* ≤
  (idealLoad * destPE - (myPrefixSum - myLoad))² **then**
    destPE = destPE - 1;
  end
end

Algorithm 2: Distributed cut algorithm.

With this distributed algorithm, the load balancer can scale to much larger machine configurations than those used in this text. However, there are not many meteorological models that scale to the level of parallelism in which this strategy is worthy—many thousands of processors [251]. To the best of our knowledge, the largest meteorological model run is presented by Michalakes *et al.* [172] with 65,536 processors, including only the dynamics portion (which does not suffer from load imbalance) of the WRF model. Therefore, we used a synthetic, fixed load to test our distributed approach.

8.6.2 Diffusion-Based Load Balancer

A common distributed load balancing strategy is based on the principle of diffusion. This principle states that energy or matter flows from higher concentrations to lower concentrations, leading to an homogeneous distribution. The flow happens between contiguous regions, i.e., energy or matter flows from one region to another that is adjacent. This principle can be applied to load balancing, so that load moves from overloaded processors to underloaded ones in the same manner. In this strategy, the processors are only required to communicate with their neighbors. This amount of communication is smaller than that required by the Hilbert-based load balancer.

Figure 8.8(a) illustrates an initial thread distribution of a processor and its neighbors, while Figure 8.8(b) shows a possible configuration after some load balancing invocations. In this load balancer, threads migrate from one

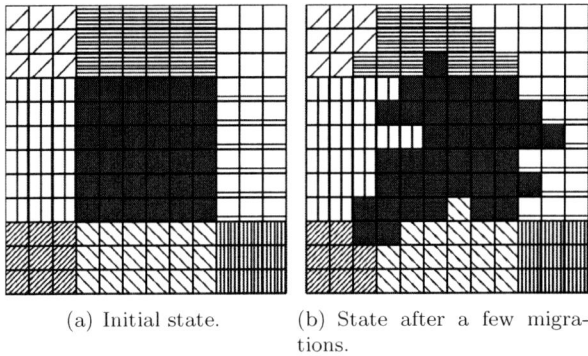

(a) Initial state.   (b) State after a few migra-
                            tions.

FIGURE 8.8: Diffusion-based load balancer.

processor to a neighbor to equalize the load between them. For each individual invocation, the load is balanced locally, but, in the long run, the whole system tends to become balanced.

For this strategy to work efficiently, some issues must be properly handled. As we stated before, neighbor threads in weather models communicate frequently, because of the exchanging boundaries. Therefore, the load balancer must take into account this natural communication when it selects a thread to migrate. The diffusion load balancer has to keep track of the threads that directly communicate with neighbor processors. This is because they are the first candidates to migrate (Figure 8.9(a)). The objective is to avoid holes that may appear in the set of threads of a processor, similarly to what happens in Figure 8.9(b).

A second issue is related with connectivity. Threads in each processor can be viewed as a directed task communication graph, where the regular communication determines the edges. The load balancer should avoid to break this graph apart, because the disconnected sub-graphs will not benefit from the local communication. One can achieve that by removing a candidate thread to migrate and running a depth-first traversal algorithm to the remaining threads. If the number of visited threads is equal to the number of remaining threads in that processor, then the candidate thread can migrate. Figure 8.9(c) has one thread that would not pass this test. Ideally, the thread graph in a single processor should be as connected as possible. In this way, the external communication is minimized.

A third issue with the diffusion-based strategy is that neighbors can move in and move out to the vicinity of a processor. Figures 8.9(d) and 8.9(e) show one example. The dark gray processor migrates some threads to the processor immediately above it. As a result, the light gray processor gains a new neighbor. This event is hard to handle, because the light gray processor does not know when a processor enters its vicinity. The incoming processor could send this information to its new neighbor, but the receiving processor does not know how long it has to wait for this message.

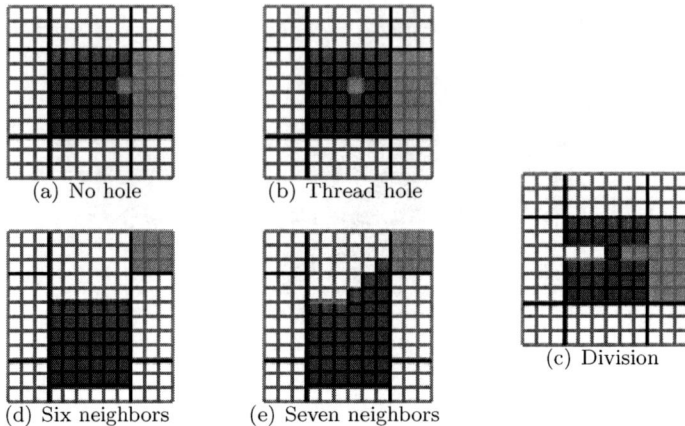

(a) No hole   (b) Thread hole

(c) Division

(d) Six neighbors   (e) Seven neighbors

FIGURE 8.9: DiffusionLB issues.

8.7   Experimental Results

This section presents experimental results. Some of these experiments were run on Kraken, a Cray-XT5 at Oak Ridge National Laboratory. Kraken has compute nodes based on AMD Opteron processors and a Seastar interconnection network with a 3D torus topology. Other experiments used Tupa, a Cray-XE6 at CPTEC-Brazil. Tupa also has AMD processors, but its network is a Gemini-based 3D torus.

We present four sets of experiments. The first one shows the performance of our TLS-based privatization strategy in comparison to the existing approach. In the second set, we consider only the virtualization effect in the application used, namely overlap of communication and computation and better cache usage. The third set of experiments compares the performance of centralized load balancers. We show that our approach produces competitive results. Finally, we analyze the performance of the fully distributed strategies in the last set of experiments.

8.7.1   First Set of Experiments: Privatization Strategy

In this section, we present performance experiments using our TLS-based privatization strategy, which was briefly described in Section 8.4.1. We used the BRAMS model to run a forecast over the southern region of Brazil. We employed a 40-level 64×64 grid and 2 Km resolution. In this experiment, we ran the model on one processor of an x86 desktop and used four different numbers of threads: 1, 2, 4 and 8. We had to replace every static variable

forecast length	1 thread	2 threads		4 threads		8 threads	
		got	tls	got	tls	got	tls
1h	509.35	497.71	491.09	505.50	476.11	554.15	474.36
6h	4241.10	4159.41	4047.24	4149.99	4005.09	4424.92	3967.77
12h	9148.00	8982.40	8860.85	9100.91	8689.95	9588.67	8719.21
24h	18583.50	18180.12	18033.24	18341.61	17631.42	19385.24	17654.10

TABLE 8.2: Execution time (seconds) of the BRAMS application.

with globals so that the original GOT-based solution could privatize them. The results are presented in Table 8.2.

In these experiments, our TLS-based privatization strategy reaches a maximum speedup of 16% over the default GOT strategy for 8 threads and 1 hour of forecast. The reason for the superior performance of our strategy is the shorter context-switch time. Since a context switch happens every time a blocking MPI call (e.g., MPI_Wait) is issued and there are many of these calls in our application, the context-switch time becomes a problem for the GOT strategy.

8.7.2 Second Set of Experiments: Virtualization Effects

In this set of experiments, we analyzed the impact of simply using processor virtualization in BRAMS, i.e., we conducted executions with different numbers of virtual processors employing a fixed number of physical processors. As the number of virtual processors increases, some overhead is expected, because over-decomposition also means that the overhead due to data movement (e.g., ghost-zone exchanges) may increase. On the other hand, a virtualized execution can benefit from automatic overlap of computation and communication, even without explicit use of non-blocking MPI calls: when a certain virtual processor blocks waiting for a message to be received, another virtual processor can proceed in the execution. Virtualization also enables better cache use, because each sub-domain is smaller than the original sub-domains in a non-virtualized scenario. Thus, these smaller sub-domains are more likely to fit in cache.

We conducted BRAMS executions under AMPI on 64 physical processors of Kraken employing 64, 256, 1024 and 2048 virtual processors. These cases corresponded to virtualization ratios of 1, 4, 16 and 32, respectively. All executions in this sub-section and in the next ones used an atmospheric grid with a total size of 512×512×40, with a 1.6 Km horizontal resolution. Table 8.3 contains the obtained results from these experiments, showing that the execution time decreases 22.4% with 256 virtual processors. With 1024 virtual processors, the decrease is slightly better, reaching 25.3%. In the case of 2048 virtual processors, however, the decrease is only 10.7%. This indicates that there must be a "sweet-spot" for the number of virtual processors to

Configuration	Wall clock time (s)
No Virtualization	4970.59
256 virtual processors	3857.53
1024 virtual processors	3713.37
2048 virtual processors	4437.50

TABLE 8.3: BRAMS execution time on 64 Kraken processors.

be used; beyond that point, using more virtual processors does not improve performance.

To understand the reasons for the performance gains when the number of virtual processors increases, we analyzed in detail the cases of 256 and 1024 virtual processors, respectively, and compared them to the non-virtualized configuration. We captured data produced by the automatic Charm++ instrumentation during the segment in the executions between time steps 1250 and 1270. This data was subsequently analyzed with the *Projections* performance analysis tool [124].

Figure 8.10 shows CPU usage for the three cases considered. In each plot, a bar corresponds to the CPU utilization on a certain physical processor during the measured segment. The first bar shows the average CPU usage, which was 44% in the execution without virtualization (Figure 8.10(a)). This usage value is low and one might suspect that the overheads associated to ghost-zones could be too high for the domain sizes being used. However, this experiment employed the typical size of a BRAMS simulation, with each sub-domain having $64 \times 64 = 4096$ columns of the atmosphere; the ghost-zones were relatively small, and contained only 256 extra columns.

In the case of four virtual processors per physical processor (Figure 8.10(b)), the average CPU usage improved to 73%. This improvement is due to the overlap between computation and communication: the waiting time that each processor would experience is now filled with computation from another virtual processor that was ready. Increasing the virtualization ratio further (i.e., to sixteen virtual processors), however, does not result in improved performance (Figure 8.10(c)), as the average CPU usage remains at 73%. This is due to the fact that the bottleneck is no longer caused by idle processors, but by load imbalance (notice that some processors in Figure 8.10(b) were already near 100% utilization). Nevertheless, this higher degree of virtualization brings the advantage of more flexibility when using migration (which becomes important for load balancing). The average CPU usage for the case of 2048 virtual processors (not presented here) was also 73%.

We also ran a second case on CPTEC's Tupa, corresponding to a different environment because its interconnection network is based on Gemini. Since this network is much faster than Seastar, the communication was not

(a) 64 processors - no virtualization

(b) 64 processors - 256 virtual processors

(c) 64 processors - 1024 virtual processors

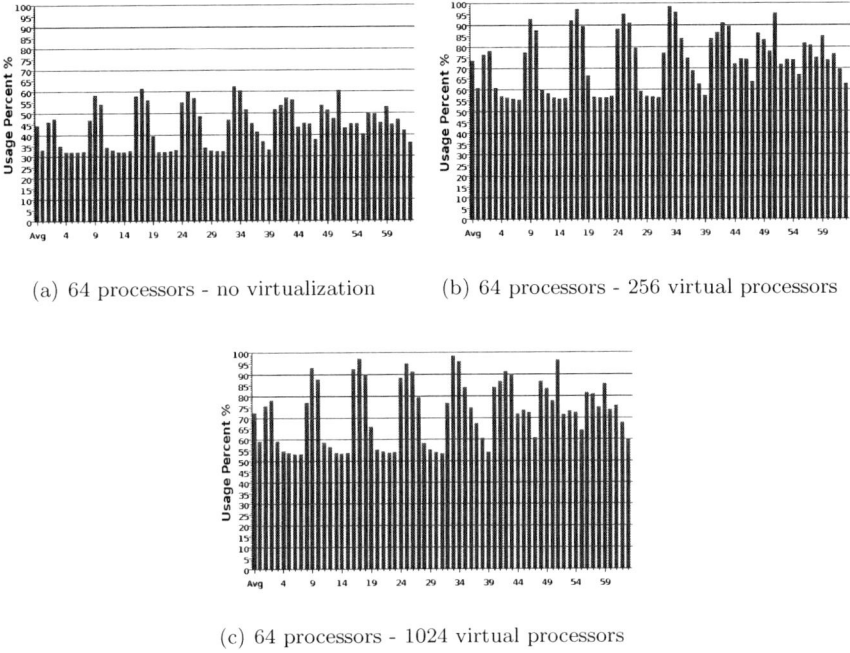

FIGURE 8.10: CPU utilization for various virtualization ratios.

a bottleneck. In this case there was no space for overlapping computation and communication. Similarly to the previous experiment, we employed 64 physical processors and 64, 256, 1024 and 2048 virtual processors. The results appear in Table 8.4. As one can see, the execution time also decreases up to a certain point in this case.

The performance improvement in this case cannot be credited to the overlapping of computation and communication. As Figure 8.11(a) shows, in the non-virtualized execution there is already one processor fully loaded; this processor holds the others, due to the intrinsic synchronization of the BRAMS execution in a time step. However, CPU usage increases with 1024 virtual pro-

Virtualization ratio	Time(s)
1	3363.4s
256	2998.8s
1024	2616.4s
2048	2677.7s

TABLE 8.4: BRAMS execution time on 64 processors of CPTEC's Tupa.

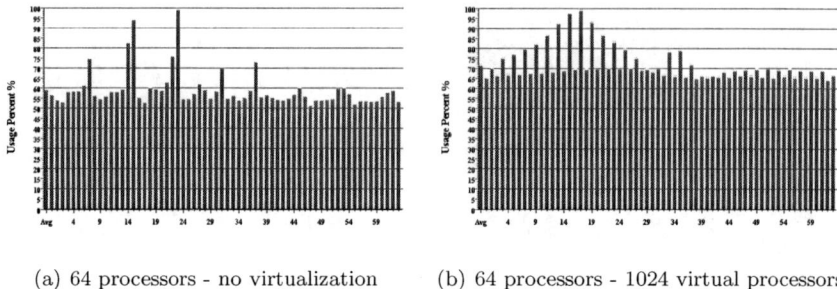

(a) 64 processors - no virtualization   (b) 64 processors - 1024 virtual processors

FIGURE 8.11: CPU utilization for a second experiment on CPTEC's Tupa.

cessors (Figure 8.11(b)). In this case, the reason for the better performance is that we employed a round-robin assignment of threads to processors. The overloaded processor in the non-virtualized execution had its load spread over several processors (each one receiving one "heavy" thread) in the virtualized execution. In this experiment we balanced load unintentionally. This result points, once more, to the need for load balancing.

We also conducted an analysis of cache utilization. This was achieved by reading the hardware performance counters of the processors via the Performance Application Programming Interface (PAPI) library [28]. Using this library in a non-virtualized environment is straightforward, but that use becomes non-trivial in a virtualized execution, because the runtime system does not guarantee that the threads in a given processor are always executed in an certain order. Therefore, we developed a scheme to ensure that the first thread entering the code section started the PAPI counters and the last thread leaving that section read those counters. Hence, our measured values for each section account for the execution of all threads on a given processor. To implement this scheme, we used global variables forced to be shared among threads.

The numbers of cache misses for the segment corresponding to the time steps of interest in the BRAMS executions are presented in Table 8.5. A consistent decrease in the number of misses can be seen in both L2 and L3 caches for the cases of 256 and 1024 virtual processors. Those results confirm the improvements in spatial locality allowed by virtualization.

The improved cache behavior observed with virtualization is due to the BRAMS code structure. As with other meteorological models, in a BRAMS time step various routines corresponding to distinct processes of nature are called in sequence (e.g., radiation, advection, etc.). Each of those routines performs its corresponding computation for the entire local sub-domain. When the sub-domain is sufficiently small to fit in cache, the first routine brings it from main memory and the remaining routines can compute with the do-

Configuration	L2 cache misses	L3 cache misses
No Virtualization	12,416M	8,448M
256 virtual processors	10,560M	4,416M
1024 virtual processors	9,408M	3,904M
2048 virtual processors	13,696M	5,056M

TABLE 8.5: Total number of cache misses on 64 processors for BRAMS.

main already in cache. Hence, all the routines (except the first one) can skip the costly phase of reading the sub-domain from main memory. This performance gain is maximized when the local sub-domain matches the size of cache.

In the case of 2048 virtual processors, as Table 8.5 shows, the number of cache misses increased. The reason for this increase, and the corresponding increase in execution time observed in Table 8.3, is that the sub-domains for this case are too small and cannot benefit from all cache space available. Because context switch among threads occurs in this virtualized execution, the data of a sub-domain remains in cache for a shorter period, being replaced by the sub-domain from another virtual processor. The same effect happens with 256 and 1024 virtual processors, but in those cases more cached data is used between context switches. We confirmed this fact by running a new experiment with 1024 virtual processors but using only 32 physical processors. The resulting numbers of L2/L3 cache misses were 9,372M and 4,038M, respectively. Those numbers are very close to the original results with a virtualization ratio of 16, despite employing a new ratio of 32. Hence, we can conclude that the lower cache utilization measured in the last row of Table 8.5 is due to the smaller sub-domain size combined with the use of virtualization. In summary, there is a "sweet-spot" in performance that is reached when the size of the sub-domain in each virtual processor best matches the underlying cache sizes, in particular the size of the L3 cache, which accounts for the most expensive misses.

8.7.3   Third Set of Experiments: Centralized Load Balancers

In addition to the benefits of cache reuse and overlapping of computation and communication presented in the previous section, the over-decomposition allows us to migrate MPI tasks to perform load balancing. In this sub-section, we evaluate centralized load balancers. We invoked such a balancer every 600 time steps during the BRAMS execution. The same grid and forecast period as before were used in this set of experiments. Table 8.6 shows the total execution time and its corresponding reduction in comparison to the base case (without virtualization).

Even though all load balancers produced better performance in comparison

Configuration	Execution Time (s)	Execution Time Reduction
No virtualization	4987.51	-
No load balancer - 1024 VP	3713.37	25.55%
GreedyLB - 1024 VP	3768.31	24.45%
RefineCommLB - 1024 VP	3714.92	25.52%
RecBisectBfLB - 1024 VP	4527.60	9.23%
MetisLB - 1024 VP	3393.12	31.97%
HilbertLB - 1024 VP	3366.99	32.50%

TABLE 8.6: Load balancing effects on BRAMS (all experiments were run on 64 real processors).

to the case without virtualization, the only ones that produced real performance gains were *HilbertLB* and *MetisLB*. The others actually lost part of the gains acquired from over-decomposition, that is, they performed worse than the "No load balancer" case. The potential reasons for these results are three: (i) the executing cost of the load balancing algorithm and the cost of migration were excessive; (ii) the cross-processor communication increased after rebalancing; and (iii) the load balancer was unable to rebalance load appropriately.

To investigate the first hypothesis, we measured the time each algorithm took to rebalance load. This result is presented in Table 8.7. Each entry in this table corresponds to the sum of the time step durations in which the load balancer was invoked; they include both the execution of the balancing algorithm itself and the thread migrations the load balancer issued. As the table shows, there is not much difference among these values (except that *RefineCommLB* was much faster, as expected, since it limits the number of migrations). Furthermore, one of the most expensive algorithms, *MetisLB*, achieved nearly the best performance. For this reason, the first hypothesis cannot explain the results we obtained.

Load Balancer	Balancing Time (s)
GreedyLB	80.81
RefineCommLB	10.81
RecBisectBfLB	78.33
MetisLB	81.00
HilbertLB	51.45

TABLE 8.7: Observed cost of load balancing.

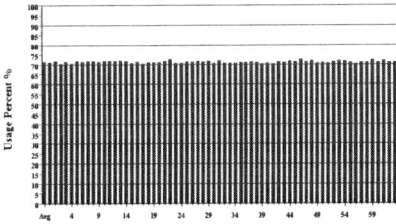

(a) GreedyLB

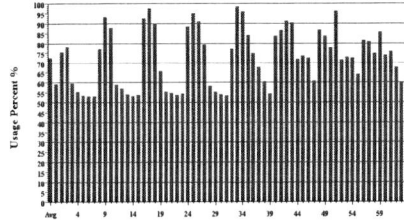

(b) RefineCommLB

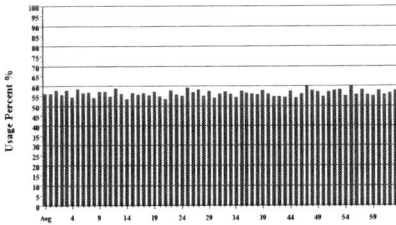

(c) RecBisectBfLB

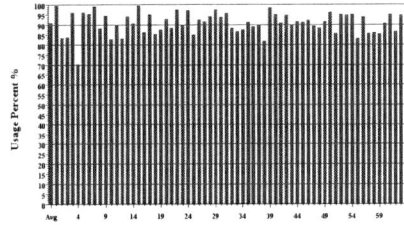

(d) MetisLB

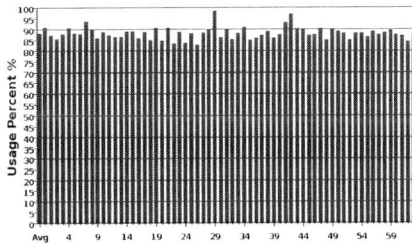

(e) HilbertLB

FIGURE 8.12: CPU usage under different load balancers.

The hypothesis of increasing cross-communication was tested firstly with the *GreedyLB* load balancer. This balancer rebalanced load quite well, as shown in Figure 8.12(a). This figure presents CPU usage of each physical processor and the first bar is the average CPU usage. We noticed that the load is well balanced but the CPU usage is low, with an average near 70%. We compared the cross-processor communication in this experiment with the one from the "No load balancer" case. We found that the cross-processor communication volume increased by a factor of nearly five with *GreedyLB* (Figure 8.13). This explains the inferior performance of *GreedyLB*.

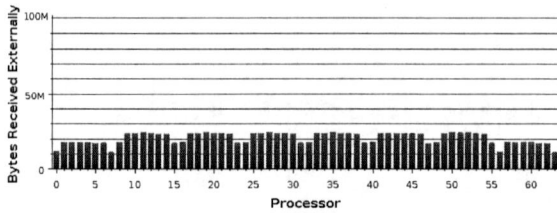

(a) No load balancer

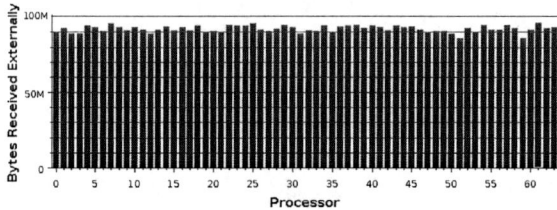

(b) *GreedyLB* balancer

FIGURE 8.13: Cross-processor communication volume.

In turn, *RefineCommLB* kept the amount of communication similar to the non-virtualized case. However, it did not migrate enough MPI tasks to completely rebalance load. As a consequence, the load was still imbalanced after its invocation (Figure 8.12(b)). That is because it assumes the load is almost balanced and it will just perform a refinement, as its name suggests. The imbalance of our experiment, however, was larger than what *RefineCommLB* could effectively handle.

Similarly to what happened to *GreedyLB*, *RecBisectBfLB* achieved good load balance (Figure 8.12(c)) but also had low CPU utilization (with an average near 55%) due to larger volume of external communication. Because the execution in BRAMS proceeds with an implicit synchronization caused by the exchange of boundary data between sub-domains at each time step, delays in one processor slow down the entire execution. We conducted additional checks and confirmed that this was indeed the cause for the poor performance of *RecBisectBfLB* observed in Table 8.6.

Finally, for *MetisLB* and *HilbertLB*, which achieved the best performance in Table 8.6, the balance was good and the average utilization was high; Figure 8.12(d) and Figure 8.12(e) show those details for those two balancers.

8.7.4   Fourth Set of Experiments: Distributed Load Balancers

So far, we have dealt exclusively with centralized load balancers. This design has the advantage that one central entity has all the information needed to take the best load balancing decision. However, a major drawback is scalability. That is because all threads have to send their observed loads to the load balancer; the load balancer has to compute the destination of each thread and it sends that decision to the appropriate thread. All these actions may represent a bottleneck as the number of processors increases.

In Section 8.6, we described a distributed load balancer that takes advantage of the regular communication pattern typically found in meteorological models. In fact, the distributed load balancer only cuts, in a distributed way, the Hilbert curve. This curve naturally keeps neighbor threads close together. In this sub-section, we compare this strategy with the centralized strategy and with a distributed diffusion based approach.

8.7.4.1 Centralized versus Distributed Load Balancers

First, we compare the balancing time of the centralized and distributed load balancers with different number of threads. Figure 8.14 shows this comparison. For a small number of threads, the centralized load balancer is fast enough, since the load balancer is invoked only a few times for the whole execution. However, its execution time increases very rapidly as the number of threads goes up. Conversely, the distributed algorithm has much shorter execution time. Even for 120,000 threads its execution time is under one-half of a second.

To analyze the reasons behind these results, we can decompose the exe-

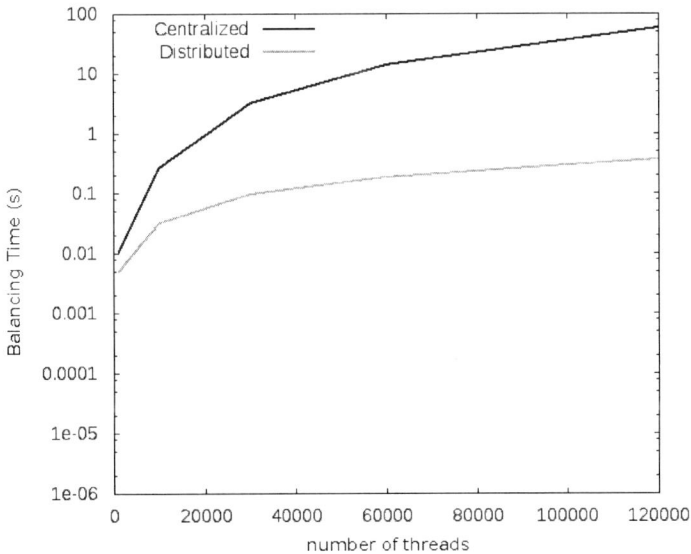

FIGURE 8.14: Comparison between the centralized and distributed algorithms.

cution time of both approaches in their components. For the centralized load balancer, the components are four: (1) the initialization phase, which corresponds to setting up the required buffers, but only needs to be done once; (2) the gathering of load in which all threads send their loads to the load balancer; (3) the algorithm that cuts the Hilbert curve according to the received loads; and (4) the scatter that sends the migration decision to each thread. The distributed load balancer has also four components: (1) the initialization phase; (2) the computation of the prefix sum, which is done by a recursive doubling algorithm, as illustrated in Figure 8.7; (3) a broadcast of the total load in the system, done by the latter processor in the Hilbert curve, which holds this value as a result of the prefix sum phase; and (4) the computation that each thread does to find its destination (Algorithm 2).

Figure 8.15 presents the time spent in each of the components of the centralized load balancer as the number of threads increases. The time of the initialization phase increases slightly, but stabilizes with 30,000 threads. This behavior is related to the way the *glibc* implements memory allocation; it takes more time to allocate large blocks, but that goes up to a certain limit [148]. Nonetheless, this initialization is quite fast and it is done only once for the whole execution. The next component is the gathering of load indexes. This communication is proportional to $\log p$, which is identical to the complexity of a reduction. However, the sizes of the messages are different. For a

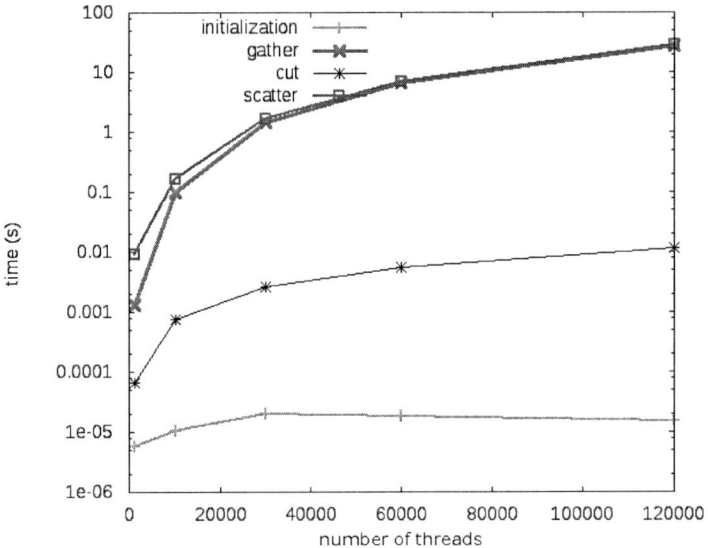

FIGURE 8.15: Execution time of each component of the centralized load balancer.

gather, the sizes increase as the algorithm executes and that makes the communication more costly than a reduction [141]. This behavior is similar to the scatter component, which has also a complexity of $\log p$ and is similar to a broadcast. These two components are the most expensive phases of this load balancer. Finally, the cut component corresponds to the segmentation of the Hilbert curve. It also scales poorly, because only one processor performs this computation.

In contrast, the execution time of the components of the distributed strategy is shown in Figure 8.16. Here, the initialization phase is very fast, because none of the threads needs to store much data. The recursive doubling part and the broadcast scale in an identical way, since both have the same complexity and the message sizes are the same, i.e., only messages of one double precision number. They represent almost all the time spent in this load balancer. Finally, the cut algorithm is also very fast, since it does not require any communication. In conclusion, the distributed strategy is faster than the centralized one because the communication in the latter scales poorly. Nonetheless, the communication in the distributed load balancer still increases when the number of processors goes up. In the next subsection, we compare this load balancer with another that has an almost constant communication time.

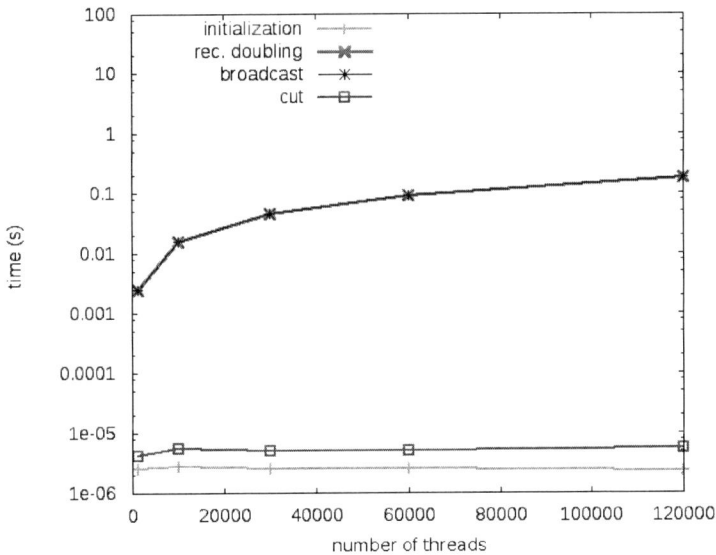

FIGURE 8.16: Execution time of each component of the distributed load balancer.

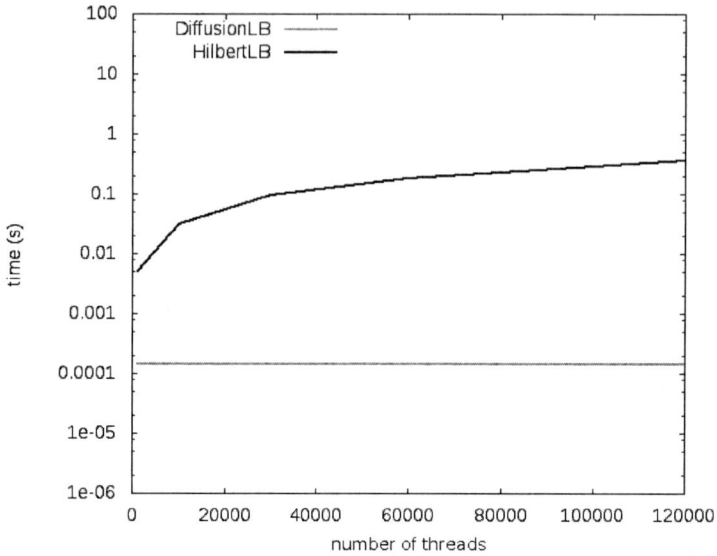

FIGURE 8.17: Execution time of two distributed load balancing approaches.

8.7.4.2   DiffusionLB versus HilbertLB

All the issues of the diffusion-based load balancer (presented in Section 8.6.2) add to the complexity of this load balancer. However, the execution time of this strategy is short as a result of the reduced amount of communication. Figure 8.17 compares the execution time of the diffusion-based and the Hilbert curve-based strategies. The former presents a constant execution time, because the number of neighbors remains approximately constant even when the total number of threads increases. Meanwhile, as shown before, the communication of the Hilbert-based strategy is proportional to $log(p)$.

Our implementation of the diffusion strategy uses a depth-first traversal algorithm to analyze the loads of threads in a given processor. This algorithm tends to keep together the threads of that processor. This approach is simpler to implement and effectively avoids breaking the graph of local threads. However, it may not produce the most cohesive graph. Nonetheless, this metric is quite similar to that observed with the Hilbert-curve strategy, as it can be seen in Figure 8.18, which shows the processors with the minimum, average and maximum number of neighbor processors after invoking the load balancer 50 times. Naturally, the diffusion-based strategy can produce the same result as the Hilbert curve, while the opposite is not true. Moreover, the former strategy has potential to optimize the communication further, because it has more degrees of freedom to select threads to migrate. Nonetheless, implementing an efficient strategy is difficult.

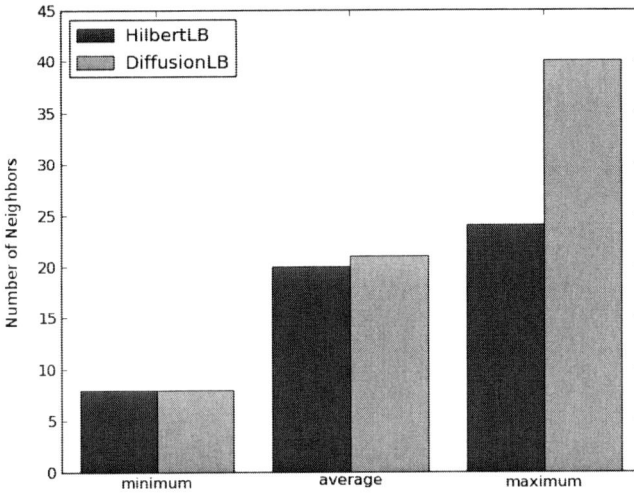

FIGURE 8.18: Number of neighbor processors after 50 invocations of the load balancer.

Although complexity is an issue for the diffusion-based load balancer, the use of virtualization isolates that from the application; hence, this problem becomes less important. However, a more critical issue is the speed of balancing of this load balancer. For each invocation, the diffusion-based strategy only balances load locally. As the time goes by, the whole system tends to became balanced. However, that may take many invocations to happen. Conversely, the Hilbert load balancer only needs one invocation to balance load completely, even though that balance may not be optimal.

Figure 8.19 shows a comparison of the speed of balancing between the diffusion-based and the Hilbert curve-based load balancers. For this experiment, we used a fixed load distribution, which represents a localized thunderstorm. We ran both load balancers for a certain number of invocations (shown in the x axis). The load in this figure is the load for the most loaded processor expressed as a percentage of its load at the beginning of the execution. The total number of threads is 16,384 and the virtualization ratio is 16.

Since the load is fixed, the Hilbert-based strategy reaches its best load distribution in the first invocation. On the other hand, the diffusion-based load balancer evolves much slower. In fact, the diffusion strategy may not even reach the same degree of balance as the Hilbert strategy. In Figure 8.19, even after 50 invocations, the load of the DiffusionLB is substantially higher than of the HilbertLB. The explanation for this fact relies on the way load moves.

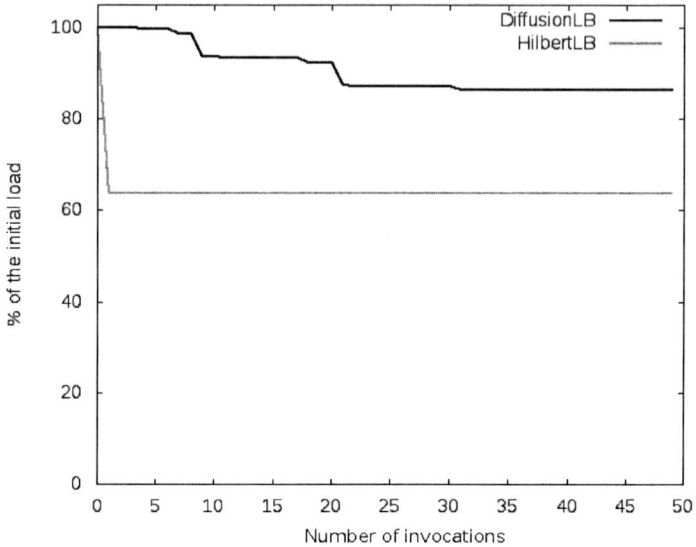

FIGURE 8.19: Balancing speed with 16K threads and virtualization ratio of 16.

In a natural diffusion process, like heat spreading throughout a medium, the energy or matter moves in a continuous manner. Meanwhile, the diffusion of load in a virtualized environment occurs in a discrete way; at least one thread has to migrate so that diffusion happens. As a consequence, the load may became "trapped." One example of this fact can be seen in Figure 8.20, in which $P0$, $P1$ and $P2$ are neighbors in this order in a 1D domain, and each processor has three threads with different loads (represented by the height of the stacked rectangles). $P0$ cannot give one thread to $P1$, because that would make the load of both much farther from average. In its turn, $P1$ does not move any load because it is already well balanced with respect to its neighbors. Therefore, in this case no diffusion occurs. Ideally, we should move only the excess of load (a fraction of one thread) from each processor to keep the load moving to homogeneity.

Increasing the virtualization ratio minimizes the possibility of load to become locked. This can be confirmed by an experiment (Figure 8.21) similar to the previous one, in which 16,384 threads execute with virtualization ratio of 64; in this way, the load becomes more "fluid" than in the previous experiment. As it can be seen, the load moves more rapidly and the system reaches stability much closer to the HilbertLB result. However, the virtualization ratio cannot be increased indefinitely. As a result, the system may never behave like a real diffusion process, in which energy or matter flows continuously from high to low concentrations.

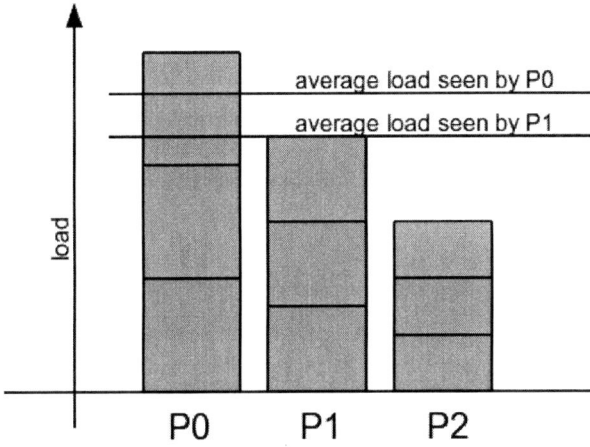

FIGURE 8.20: Load "trapped."

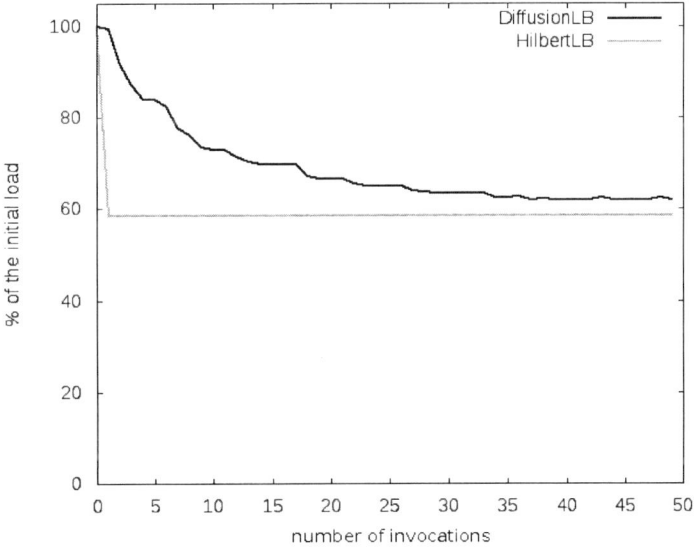

FIGURE 8.21: Balancing speed with 16K threads and virtualization ratio of 64.

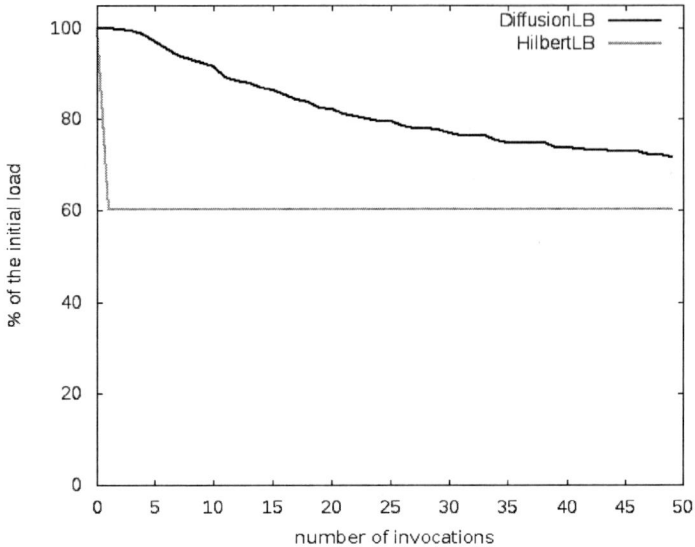

FIGURE 8.22: Balancing speed with 64K threads and virtualization ratio of 64.

Although we can increase the virtualization ratio (up to a certain point), increasing the total number of processors (real processors) slows down the speed of balance of the diffusion-based strategy. If we run the same experiment as before (virtualization of 64) with more processors (1024), the system takes much longer to become balanced. Figure 8.22 shows the result of this experiment.

In summary, although the diffusion-based load balancer is distributed and has potential to optimize the communication even better than the Hilbert curve-based load balancer, the complexity and, more important, the speed of balance make this strategy inappropriate to the application considered.

8.8   Final Remarks

Load imbalance is a major impediment to scalability. It refers to the uneven distribution of load across tasks of a parallel application. The most loaded task dominates application completion time. If the application is synchronous, the most loaded task also delays the execution of the other tasks. Ideally, the parallel application should keep load equally distributed among its tasks, so that the efficiency is maximized.

Many load balance strategies have been developed. Nonetheless, many real-world applications lack such feature. This is particularly true for meteorological models. Although these applications typically use a regular three-dimensional grid and, in a parallel execution, each processor executes the same code, meteorological models are subject to load imbalance. Xue *et al.* [247] declare that processors in charge of active thunderstorms may incur up to 30% additional computation. With the increase in complexity, this imbalance may grow even larger. Most models currently do not perform load balance, due to the complexity involved with this task.

The difficulty of applying load balancing strategies to real weather models stems from the application complexity. For example, the BRAMS model, used in this work, has 140K lines of code and a few thousand routines. Similarly, other models, such as WRF, are also large and complex. These applications are developed by many different programmers during long periods. Moreover, they are typically designed without any concern about load imbalance.

In this text, we present a strategy to perform load balancing. This strategy is based on processor virtualization and its implementation in AMPI. Its main advantage is simplicity; fewer changes to the original application are required than those needed by embedded load balancers. This work is the first one to successfully use processor virtualization to balance load in a complete meteorological model. Previous attempts were either too slow or not able to run more than one task per processor, which is allowed by the processor virtualization technique.

To use processor virtualization, the user has to over-decompose the parallel application in more tasks than processors. Then, a set of tasks, also known as Virtual Processors, is assigned to each processor. If a processor becomes overloaded, then some of its tasks can migrate to underloaded processors. In our experiments, we found that just the over-decomposition is already beneficial for the application: we obtained up to 25.5% reduction in execution time [206]. This reduction can be attributed to the overlap of communication and computation and better use of the cache hierarchy.

AMPI implements virtual processors as user-threads. Its main advantage is that user-threads are very lightweight. However, this approach changes the semantic that is usually assumed by MPI programs. That is because AMPI tasks running in a same processor do not have a private address space; the static and global variables are shared among those tasks. We developed a new strategy to privatize global and static data so that an existing application can be more easily ported to the AMPI environment. This strategy has a better context switch time than a previous privatization scheme [204].

Using AMPI, however, is not enough to deal with the load balancing issues of meteorological models or some other applications. A load balancing strategy has to be employed. It is responsible for mapping threads to processors whenever the load balancer is invoked. We explored some existing load balancers that, in principle, would fit the meteorological model requirements. We also developed a Hilbert curve-based load balancer. It maps the 2D domain decomposition to a 1D space. The curve is then cut into segments so that each segment has approximately the same load. Due to properties of the Hilbert curve, the corresponding sub-domains on a given segment should be close in the 2D space and, consequently, the cross-processor communication is reduced. This load balancer reduced the execution time by 7% [205] beyond what had already been obtained by virtualization alone.

The previously mentioned load balancers were centralized. They have a single processor to gather load balancing information and take load balancing decisions. This approach works well for moderately large machines. However, for larger machines, fully distributed load balancers are more appropriate. Typically, these algorithms are diffusion-based, i.e., the load "flows" from higher to lower concentrations. In this way, no processor needs to know the load distribution of the entire machine; they need only the load of their neighbors. However, this strategy tends to be very slow to converge.

Our centralized load balancing algorithm can become distributed in a way that each processor still has information about the whole system, but not the individual loads of each processor. This alleviates the communication requirements. In this text, we compared the centralized and distributed versions of the Hilbert curve-based load balancer. The centralized version does not scale well, as expected, and the reasons for that are the gathering of load indexes and the scattering of load balancing decisions. In its turn, the communication of the distributed version scales better, but it is still has a cost proportional to $log(p)$. We then compared the distributed approach with a diffusion-based strategy, because the latter has a more efficient communication. However, as shown in our results, the diffusion-based load balancer takes much longer to balance load. Furthermore, since the load cannot move continuously but rather in discrete amounts (the total load of a thread), some imbalance may get locked.

Finally, although our strategy has been developed for weather forecast models, it can also be applied to other applications, as long as the same structure and behavior is present (for example [242]). Our Hilbert curve load balancer depends on the regularity of the communication pattern; threads communicate with neighbors forming a Cartesian grid. That can be extended provided that a new space-filling curve that fits the new communication pattern is used. In addition, the behavior of the application must be similar to weather forecast models, in the sense that there is no massive imbalance. We expect that applications such as Fluid Dynamics and Ocean Modeling can benefit from our strategy.

Acknowledgments

This work was supported by grants from Brazil's National Council for Scientific and Technological Development (CNPq) and the U.S. Department of Energy (DE-SC0001845). The authors are also grateful to Philippe O. A. Navaux and Alvaro Fazenda for their invaluable help.

This work used the Extreme Science and Engineering Discovery Environment (XSEDE), which is supported by National Science Foundation grant number OCI-1053575 (project allocations TG-ASC050039N and TG-ASC050040N).

Chapter 9

Crack Propagation Analysis with Automatic Load Balancing

Orion Sky Lawlor

Department of Computer Science, University of Alaska Fairbanks

M. Scot Breitenfeld and Philippe H. Geubelle

Department of Aerospace Engineering, University of Illinois at Urbana-Champaign

Gengbin Zheng

National Center for Supercomputing Applications, University of Illinois at Urbana-Champaign

9.1 Introduction

Researchers in the field of structural mechanics often turn to the parallel finite element method to model physical phenomena with finer detail, sophistication and accuracy. While parallel computing can provide large amounts of computational power, developing parallel software requires substantial effort to exploit this power efficiently.

A wide variety of applications involve explicit computations on unstructured grids. One of the key objectives of this work is to create a flexible framework to perform these types of simulations on parallel computing platforms. Parallel programming introduces several complications:

- Simply expressing a computation in parallel requires the use of either a specialized language such as HPF [134] or an additional library such as MPI.

- Parallel execution makes race conditions and nondeterministic execution possible. Some languages, such as HPF, have a simple lockstep control structure and are thus relatively immune to this problem. However, in other implementations, such as asynchronous MPI or threads, these issues are more common.

- Computation and communication must be overlapped to achieve optimal performance. However, few languages provide good support for this overlap, and even simple static schemes can be painfully difficult to implement in the general case, such as overlapping computation from one library with communication from a different library.

- Load imbalance can severely restrict performance, especially for dynamic applications. Automatic or application-independent load balancing capabilities are rare (Section 9.2.2).

Our approach to managing the complexity of parallel programming is based on a simple division of labor. In this approach, parallel programming specialists in computer science provide a simple but efficient *parallel framework* for the computation, while application specialists provide the numerics and physics. The parallel framework described hereafter abstracts away the details of its parallel implementation.

Since the parallel framework is application independent, it can be reused across multiple codes. This reuse amortizes the effort and time spent developing the framework and makes it feasible to invest in sophisticated capabilities such as adaptive computation and communication overlap and automatic measurement-based load balancing. Overall, this approach has proven quite effective, leveraging skills in both computer science and engineering to solve problems neither could solve independently.

9.1.1   ParFUM Framework

ParFUM [144] is a parallel framework for performing explicit computations on unstructured grids. The framework has been used for finite-element computations, solving partial differential equations, computational fluid dynamics and other problems.

The basic abstraction provided is very simple—the computational domain consists of an irregular mesh of nodes and elements. The elements are divided into partitions or chunks, normally using the graph partitioning library METIS [72], or ParMETIS [71]. These chunks reside in migratable AMPI virtual processors, thereby taking advantage of runtime optimizations including dynamic load balancing. The chunks of meshes and AMPI virtual processors are then distributed across the processors of the parallel machine. There is typically at least one chunk per processor, but usually many more. Mesh nodes can be either private, adjacent to the elements of a single partition, or shared, adjacent to the elements of different partitions.

A ParFUM application has two main subroutines: *init* and *driver*. The *init* subroutine executes only on processor 0 and is used to read the input mesh and physical data and register it with the framework. The framework then partitions the mesh into as many regions as requested, each partition being a virtual processor. It then executes the *driver* routine on each virtual processor. This routine computes the solution over the local partition of the mesh.

The solution loop for most applications involves a calculation in which each mesh node or element requires data from its neighboring entities. Thus entities on the boundary of a partition need data from entities on other partitions. ParFUM provides a flexible and scalable approach to meet an application's communication requirements. ParFUM adds local read-only copies of remote entities to the partition boundary. These read-only copies are referred to as *ghosts*. A single collective call to ParFUM allows the user to update all ghost mesh entities with data from the original copies on neighboring partitions. This lets application code have effortless access to data from neighboring entities on other partitions. Since the definition of "neighboring" can vary from one application to another, ParFUM provides a flexible mechanism for generating ghost layers. For example, an application might consider two tetrahedra that share a face as neighbors. In another application, tetrahedra that share edges might be considered neighbors. ParFUM users can specify the type of ghost layer required by defining the "neighboring" relationship in the *init* routine and adding multiple layers of ghosts according to the neighboring relationship for applications that require them. In addition, the definition of "neighboring" can vary for different layers. User-specified ghost layers are automatically added after partitioning the input mesh provided during the *init* routine. ParFUM also updates the connectivity and adjacency information of a partition's entities to reflect the additional layers of ghosts. Thus ParFUM satisfies the communication needs of a wide range of applications by allowing the user to add arbitrary ghost layers. After the communication for ghost layers, each local partition is nearly self-contained; a serial numerics routine can be run on the partition with only a minor modification to the boundary conditions.

With the above design, the ParFUM framework enables straightforward conversion of serial codes into parallel applications. For example, in an explicit structural dynamics computation, each iteration of the time loop has the following structure:

1. Compute element strains based on nodal displacements.

2. Compute element stresses based on element strains.

3. Compute nodal forces based on element stresses.

4. Apply external boundary conditions.

5. Update nodal accelerations, velocities and displacements based on Newtonian physics.

In a serial code, these operations apply over the entire mesh. However, since each operation is local, depending only on a node or element's immediate neighbors, we can partition the mesh and run the same code on each partition.

The only problem is ensuring that the boundary conditions of the different partitions match. We might choose to duplicate the nodes along the boundary and then sum up the nodal forces during step 3, which amounts to the simple change in step 4 to: Apply external *and internal* boundary conditions.

For existing codes that have already been parallelized with MPI, the conversion to ParFUM is even faster, thereby taking advantage of Charm++ features, such as dynamic load balancing.

9.1.2   Implementation of the ParFUM Framework

As shown in the software architecture diagram of Figure 9.1, our parallel FEM framework is written and parallelized using Adaptive MPI. Adaptive MPI (AMPI) [106, 105] is an MPI implementation and extension based on the Charm++ [117] programming model. Charm++ is a parallel C++ runtime system that embodies the concept of *processor virtualization* [122]. The idea behind processor virtualization is that the programmer decomposes the computation, without regard to the physical number of processors available, into a large number of logical work units and data units, which are encapsulated in *virtual processors* (VPs) [122]. The programmer leaves the assignment of VPs to physical processors to the runtime system, which incorporates intelligent optimization strategies and automatic runtime adaptation. These virtual processors themselves can be programmed using any programming paradigm. They can be organized as indexed collections of C++ objects that interact via asynchronous method invocations, as in Charm++ [147]. Alternatively, they can be MPI virtual processors implemented as user-level, lightweight threads (not to be confused with system-level threads or Pthreads) that interact with each other via messages, as in AMPI (illustrated in Figure 9.2).

This idea of processor virtualization brings significant benefits to both parallel programming productivity and parallel performance [123]. It empowers the runtime system to incorporate intelligent optimization strategies and automatic runtime adaptation. The following is a list of the benefits of processor virtualization.

Automatic load balancing: AMPI threads (the virtual processors) are decoupled from real processors. Therefore, they are location independent and

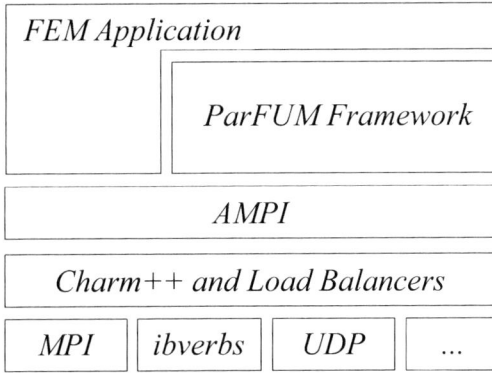

FIGURE 9.1: Software layer diagram for finite element codes in Charm++.

can migrate from processor to processor. Thread migration provides the basic mechanism for load balancing: if some of the physical processors become overloaded, the runtime system can migrate a few of their AMPI threads to underloaded physical processors. The AMPI runtime system provides transparent support of message forwarding after thread migration.

Adaptive overlapping of communication and computation: If one of the AMPI threads is blocked on a receive, another AMPI thread on the same physical processor can run. This largely eliminates the need for the programmer to manually specify some static computation/communication overlapping, as is often required in MPI.

Optimized communication library support: Besides the communication optimization inherited from Charm++, AMPI supports asynchronous or nonblocking interfaces to collective communication operations. This allows the overlapping between time-consuming collective operations and other useful computation.

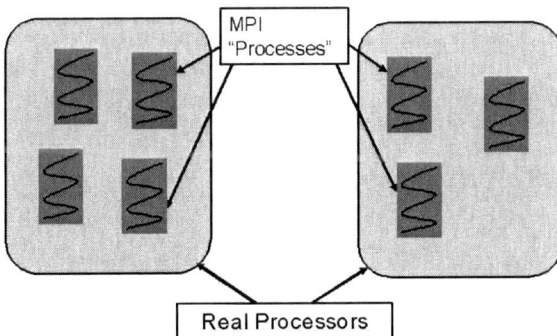

FIGURE 9.2: Implementation of AMPI virtual processors.

Better cache performance: A virtual processor handles a smaller set of data than a physical processor, so a virtual processor will have better memory locality. This blocking effect is the same method many *serial* cache optimizations employ, and AMPI programs get this benefit automatically.

Flexibility to run on an arbitrary number of processors: Since more than one VP can be executed on one physical processor, AMPI is capable of running MPI programs on any arbitrary number of processors. This feature proves to be useful in application development and debugging phases.

In many applications, we have demonstrated that the processor virtualization does not incur much cost in parallel performance [123], due to low scheduling overheads of user-level threads. In fact, it often improves cache performance significantly because of its blocking effect.

Charm++ and AMPI have been used as mature parallelization tools and runtime systems for a variety of real world applications for scalability [197, 140, 239, 80]. With these successes of improving parallel efficiency, several domain-specific frameworks on top of Charm++ and AMPI have been developed to further enhance programmer productivity while automating the parallelization process, which produces reusable libraries for parallel algorithms.

Additionally, Charm++ supports several useful features for monitoring running applications. Converse Client Server (CCS) [113, 47] provides a socket-based transport layer to get data in and out of any program, and NetFEM provides a higher level nodes-and-elements interface for remote online visualization, as shown in Figure 9.3.

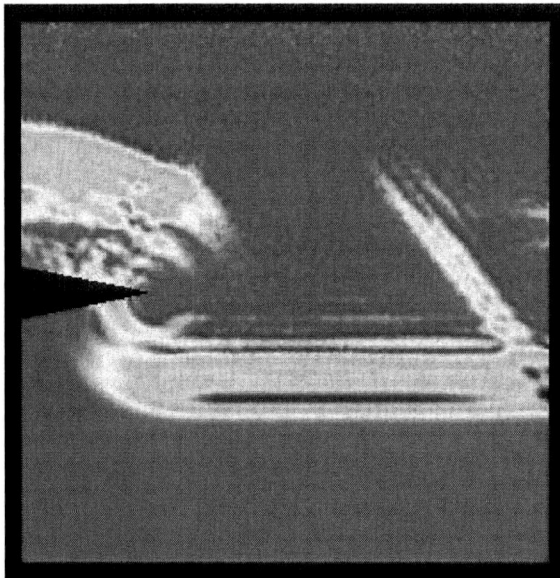

FIGURE 9.3: Diffraction off a crack, simulated in parallel and rendered using NetFEM directly from the running simulation (**see Color Plate 7**).

9.2   Load Balancing Finite Element Codes in Charm++

Among the challenges associated with the parallelization of finite element codes, achieving load balance is key to scaling a dynamic application to a large number of processors. This is especially true for dynamic structural mechanics codes where simulations involve rapidly evolving geometry and physics, often resulting in a load imbalance between processors. As a result of this load imbalance, the application has to run at the speed of the slowest processor with deteriorated performance. The load imbalance problem has driven decades of research activities in load balancing techniques [30, 6, 251, 7].

Of interest in this work is dynamic load balancing, which attempts to solve the load balance problem at runtime according to the most up-to-date load information. This approach is a challenging software design issue and generally creates a burden for the application developers, who often must include the mechanism to inform the decision-making module concerning load balance the estimated CPU load and the communication structure. In addition, once load imbalance is detected and data migration is requested, a developer has to write complicated code for moving data across processors. The ideal load balancing framework should hide the details of load balancing so that the application developer can concentrate on modeling the physics of the problem.

In this section, we present an automatic load balancing method and its application to the three-dimensional finite element simulations of wave propagation and dynamic crack propagation events. The parallelization model used in this application is the processor virtualization supported by the migratable MPI threads. The application runs on a large number of MPI threads (that exceeds the actual physical number of processors), allowing to perform runtime load balancing by migrating MPI threads. The MPI runtime system automatically collects load information from the execution of the application. Based on this instrumented load data, the runtime module makes decisions on migrating MPI threads from heavily loaded processors to underloaded ones. This approach thus requires minimal efforts from the application developer.

9.2.1   Runtime Support for Thread Migration

In ParFUM applications, load balancing is achieved by migrating AMPI threads that host mesh partitions from overloaded processors to underloaded ones. When an AMPI thread migrates between processors, it must move all the associated data, including its stack and heap-allocated data. The Charm++ runtime supports both fully automated thread migration [253] and flexible user-controlled migration of data by additional helper functions.

In fully automatic mode, the AMPI runtime system automatically transfers a thread's stack and heap data which are allocated by a special memory allocator called an *isomalloc* [105], in a manner similar to that of *PM2* [6]. It is portable on most platforms except for those where the `mmap` system call is unavailable. Isomalloc allocates data with a globally unique virtual address, reserving the same virtual space on all processors. With this mechanism, iso-malloced data can be moved to a new processor without changing the address. This provides a clean way to move a thread's stack and heap data to a new machine automatically. In this case, migration is transparent to the user code.

Alternatively, users can write their own helper functions to pack and un-pack heap data for migrating an AMPI thread. This is useful when application developers wish to have more control in reducing the data volume by using application specific knowledge and/or by packing only variables that are live at the time of migration. The PUP (Pack/UnPack) library [113] was written to simplify this process and reduce the amount of code the developers have to write. The developers only need to write a single PUP routine to traverse the data structure and this routine is used for both packing and unpacking.

9.2.2   Comparison to Prior Work

The goal of our work is a generic load balancing framework that optimizes the load balance of the irregular and highly dynamic applications with an application independent interface; therefore, we will focus our discussion in this section to those dynamic load balancing systems for parallel applications. In particular, we wish to distinguish our research using the following criteria:

- Supporting data migration. Migrating data has advantages over migrat-ing "heavy-weight" processes, which adds complexity to the runtime system.

- General Purpose. Load balancing methods are designed to be application independent. They can be used for a wide variety of applications.

- Communication-aware load balancing. The framework takes communi-cation into account explicitly, unlike implicit schemes which rely on domain-specific knowledge. Communication patterns, including multi-cast relationships and communication volume, are directly recorded into a load balancing database for load balancing algorithms.

- Automatic load measurement. The load balancing framework does not rely on the application developer to provide application load informa-tion, but measures computational costs at runtime.

- Adaptive to execution environment. Takes background load and non-migratable load into account.

Table 9.1 shows the comparison of the Charm++ load balancing frame-work to several other software systems that support dynamic load balancing. DRAMA [16] is designed specifically to support finite element applications. This specialization enables DRAMA to provide an "application independent"

System Name	Data Migration	General Purpose	Network Aware	Automatic Measurement	Adaptive
DRAMA	Yes	No	No	Yes	No
Zoltan	Yes	Yes	No	No	No
PREMA	Yes	No	No	No	No
Chombo	Yes	No	No	No	No
Charm++	Yes	Yes	Yes	Yes	Yes

TABLE 9.1: Software systems that support dynamic load balancing.

load balancing using its built-in cost functions for the category of applications. Zoltan [49, 50] does not make assumptions about applications' data and is designed to be a general purpose load balancing library. However, it relies on application developers to provide a cost function and communication graph. The system PREMA [8, 9] supports a very similar idea of migratable objects. However, its load balancing method primarily focuses on task scheduling problems as in noniterative applications. The Chombo [40] package has been developed by Lawrence Berkeley National Lab. It provides a set of tools including load balancing for implementing finite difference methods for the solution of partial differential equations on block-structured adaptively refined rectangular grids. It requires users to provide input for computational workload as a real number for each box (defined as a partition of mesh). Charm++ provides the most comprehensive features for load balancing. The measurement-based load balancing scheme enables automatic adaptation to application behavior. It is applicable to most scientific and engineering applications where computational load is persistent, even if it is dynamic. Charm++ load balancing is also capable of adapting to the change of background load [29] of the execution environment.

9.2.3   Automatic Load Balancing for FEM

Many modern explicit finite element applications are used to solve highly unsteady, dynamic, irregular problems. For example, an elasto-plastic solid mechanics simulator that we explore in Section 9.3 might use a different force calculation for highly stressed elements undergoing plastic deformation. Another simulation might use dynamic geometric mesh refinement to follow dynamic shocks [160]. In these applications, load balancing is required to achieve the desired high performance on large parallel machines.

ParFUM directly utilizes the load balancing framework in Charm++ and AMPI [251, 20]. The load balancing involves four distinct steps: (1) load evaluation; (2) load balancing initiation to determine when to start a new load balancing process; (3) load balancing decision making and (4) task and data migration.

The Charm++ load balancing framework adopts a unique measurement-based strategy for load evaluation. This scheme is based on runtime instrumentation, which is feasible due to the *principle of persistence* that can be found in most physical simulations: the communication patterns between objects as well as the computational load of each of them tend to persist over time, even in the case of dynamic applications. This implies that the recent past behavior of a system can be used as a good predictor of the near future. The load instrumentation is fully automatic at runtime. During the execution of a ParFUM application, the runtime measures the computation load for each object and records communication pattern into a load "database" on each processor. This approach provides an automatic load balancing solution that can adapt to application behavior while requiring minimal effort from the developers.

The runtime then assesses the load database periodically and determines if load imbalance is present. The load imbalance can be computed as:

$$\sigma = \frac{L_{max}}{L_{avg}} - 1, \tag{9.1}$$

where L_{max} is the maximum load across all processors, and L_{avg} is the average load of all the processors. Note that even when load imbalance occurs ($\sigma > 0$), it may not be profitable to start a new load balancing step due to the overhead of load balancing itself. In practice, a load imbalance threshold can be chosen based on a heuristic that the gain of the load balancing ($L_{max} - L_{avg}$) is at least greater than the estimated cost of the load balancing (C_{lb}). That is:

$$\sigma > \frac{C_{lb}}{L_{avg}}. \tag{9.2}$$

When load balancing is triggered, the load balancing decision module uses the load database to compute a new assignment of virtual processors to physical processors and informs the runtime to execute the migration decision.

9.2.4   Load Balancing Strategies

In the step that makes the load balancing decision, the Charm++ runtime assigns AMPI threads on physical processors, so as to minimize the maximum load (makespan) on the processors. This is known as the Makespan minimization problem, and the exact solution has been shown to be an NP-hard optimization problem [151]. However, many combinatorial algorithms have been developed that find a reasonably good approximate solution. Charm++ load balancing framework provides a spectrum of simple to sophisticated heuristic-based load balancing algorithms, some of which are described in more detail below:

- Greedy Strategy: This simple strategy organizes all the objects in decreasing order of their computation times. The algorithm repeatedly selects the heaviest un-assigned object and assigns it to the least loaded processor. This algorithm may lead to a large number of migrations. However, it works effectively in most cases.

- Refinement Strategy: The refinement strategy is an algorithm which improves the load balance by incrementally adjusting the existing object distribution, especially on highly loaded processors. The computational cost of this algorithm is low because only a subset of processors is examined. Furthermore, this algorithm results in only a few objects being migrated, which makes it suitable for fine-tuning the load balance.

- METIS-based Strategy: This strategy uses the METIS graph partitioning library [130] to partition the object-communication graph. The objective of this strategy is to find a reasonable load balance, while minimizing the communication among processors.

The Charm++ load balancing framework also allows a developer to implement his own load balancing strategies based on heuristics specific to the target application (such as in the NAMD [197] molecular simulation code).

Load balancing can be done in either centralized or distributed approach depending on how the load balancing decisions are made. In the centralized approach, one central processor makes the decisions globally. The load databases of all processors are collected to the central processor, which may incur high communication overhead and memory usage for the central processor. In the distributed approach, load balance decisions are made in a distributed fashion, where load data is only exchanged among neighboring processors. Due to the lack of the global information and aging of the load data, distributed load balancing tends to converge slowly to the good load balance discovered by the centralized approach. Therefore, we typically use a centralized or global load balancing strategy.

9.2.5 Agile Load Balancing

Applications with rapidly changing load require frequent load balancing, which demands rapid load balancing with minimal overhead. Normal load balancing strategies in Charm++ occur in *synchronous* mode, as shown in Figure 9.4. At load balancing time, the application on each processor stops after it finishes its designated iterations and hands control to the load balancing framework to make load balancing decisions. The application can only resume when the load balancing step finishes and all AMPI threads migrate to the destination processors. In practice, this "stop and go" load balancing scheme is simple to implement, and has one important advantage—thread migration happens under user control, so that a user can choose a convenient time for the thread migration, to minimize the implementation complexity and runtime data size of the migration. However, this scheme is not efficient due to the effect of the global barrier. It suffers from high overhead due to the fact that the load balancing process on the central processor has to wait

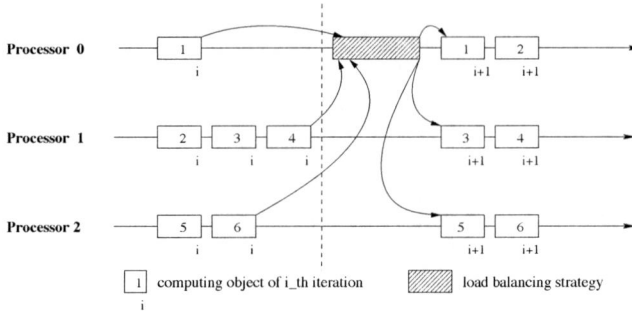

FIGURE 9.4: Traditional synchronous load balancing.

for the slowest processor to join load balancing, thus wasting CPU cycles on other processors. This motivated the development of an agile load balancing strategy that performs *asynchronous* load balancing which allows overlapping of load balancing time and normal computation.

The asynchronous load balancing scheme takes full advantage of Charm++'s intelligent runtime support for concurrent compositionality [123] that allows dynamic overlapping of the execution of different composed modules in time and space. In the asynchronous scheme, the load balancing process occurs concurrently, or in the background of normal computation. When it is time for load balancing, each processor sends its load database to the central processor and continues its normal computation without waiting for load balancing to start. When a migration decision is calculated at the background on the central processor, the AMPI threads are instructed to migrate to their new processors in the middle of their computation.

There are a few advantages of asynchronous load balancing over the synchronous scheme. First, eliminating the global barrier helps in reducing the idle time on faster processors which otherwise would have to wait for the slower processors to join the load balancing step. Second, it allows the overlapping of load balancing decision making time and computation in an application, which potentially could help improve the overall performance. Finally, each thread can have more flexible control on when to migrate to the designated processor. For example, a thread can choose to migrate when it is about to be idle, which potentially allows overlapping of the thread migration and computation of other threads.

Asynchronous load balancing, however, imposes a significant challenge to thread migration in the AMPI runtime system. AMPI threads may migrate *at any time*, whenever they receive the migration notification. In practice, it is not trivial for an AMPI thread to migrate at any time due to the complex runtime state involved, for example when a thread is suspended in the middle of pending receives. In order to support any-time migration of AMPI threads, we extended the AMPI runtime to be able to transfer a complete runtime state associated with the AMPI threads including the pending receive requests and buffered messages for future receives. With the help of isomalloc stack and heap, AMPI threads can be migrated to a new processor transparently at any time: a thread can actually be suspended on one processor, migrated and resumed on a different processor in a new address space. For AMPI threads with pending receives, incoming messages are redirected automatically to the destination processors by the runtime system.

In the next section, we present a simulation case study to demonstrate the effectiveness of our finite element framework and load balancing strategies.

9.3   Cohesive and Elasto-plastic Finite Element Model of Fracture

To simulate the spontaneous initiation and propagation of a crack in a discretized domain, an explicit cohesive-volumetric finite element (CVFE) scheme [246], [31], [74] is used. As its name indicates, the scheme relies on a combination of volumetric elements used to capture the constitutive response of the continuum medium, and cohesive interfacial elements to model the failure process taking place in the vicinity of the advancing crack front. The CVFE concept is illustrated in Figure 9.5, which presents two 4-node tetrahedral volumetric elements tied together by a 6-node cohesive element shown in its deformed configuration, as the adjacent nodes are initially superposed and the cohesive element has no volume.

In the present study, the mechanical response of the cohesive elements is described by the bilinear traction-separation law illustrated in Figure 9.6 for the case of tensile (Mode I) failure. After an initial stiffening (rising) phase, the

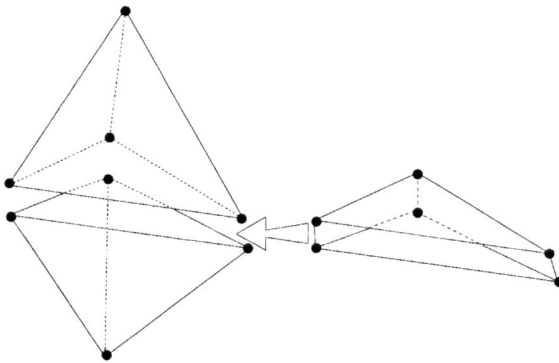

FIGURE 9.5: Two 4-node tetrahedral volumetric elements linked by a 6-node cohesive element.

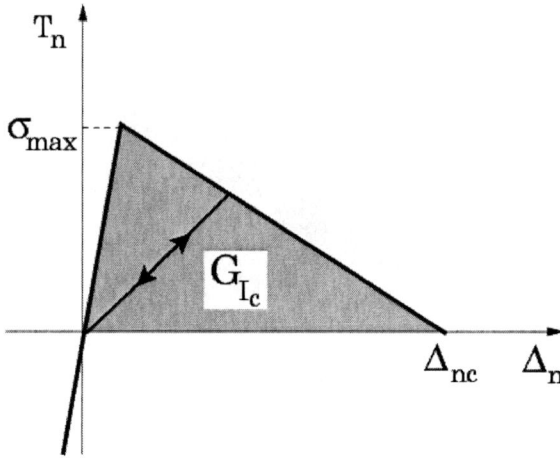

FIGURE 9.6: Bilinear traction-separation law for mode I failure modeling. The area under the curve corresponds to the mode I fracture toughness G_{I_c} of the material.

cohesive traction T_n reaches a maximum corresponding to the failure strength σ_{max} of the material, followed by a downward phase that represents the progressive failure of the material. Once the critical value Δ_{nc} of the displacement jump is reached, no more traction is exerted across the cohesive interface and a traction-free surface (i.e., a crack) is created in the discretized domain. The emphasis of the dynamic fracture study summarized hereafter is on the simulation of purely Mode I failure, although cohesive models have also been proposed for the simulation of mixed-mode fracture events. Also illustrated in Figure 9.6 is an unloading and reloading path followed by the cohesive traction during an unloading event taking place while the material fails.

The finite element formulation of the CVFE scheme is derived from the following form of the principle of virtual work:

$$\int_V (\rho \ddot{u}_i \, \delta u_i + S_{ij} \, \delta E_{ij}) \, dV = \int_{S_T} T_i^{ex} \, \delta u_i \, dS_T + \int_{S_c} T_i \, \delta \Delta_i \, dS_c, \quad (9.3)$$

where the left-hand side corresponds to the virtual work done by the inertial forces ($\rho \ddot{u}_i$) and the internal stresses (S_{ij}), and the right-hand side denotes the virtual work associated with the externally applied traction (T_i^{ex}) and cohesive traction (T_i) acting along their respective surfaces of application S_T and S_c. In Equation (9.3), ρ denotes the material density, u_i and E_{ij} are the displacement and strain fields, respectively, and Δ_i denotes the displacement jump across the cohesive surfaces. The implementation relies on an explicit time stepping scheme based on the central difference formulation [246]. A nonlinear kinematics description is used to capture the large deformation and rotation associated with the propagation of the crack. The strain measure used here is the Lagrangian strain tensor $\mathbf{E}$.

To complete the CVFE scheme, we need to model the constitutive response of the material, i.e., to describe the response of the volumetric elements. In the present study, we use an explicit elasto-visco-plastic update scheme nonlinear elasticity, which is compatible with the nonlinear kinematic description and relies on the multiplicative decomposition of the deformation gradient $\mathbf{F}$ into elastic and plastic parts as

$$\mathbf{F} = \mathbf{F}^e \mathbf{F}^p. \tag{9.4}$$

The update of the plastic component $\mathbf{F}^p$ of the deformation gradient at the $(n+1)^{th}$ time step is obtained by

$$\mathbf{F}^p_{n+1} = exp\left[\sum_A \frac{\Delta\gamma}{\sqrt{2}\tilde{\sigma}}\left(\sigma^A - \frac{I_1^\sigma}{3}\right)\mathbf{N}^A \otimes \mathbf{N}^A\right] \bullet \mathbf{F}^p_n, \tag{9.5}$$

where $\mathbf{N}^A$ (A=1, 2, 3) denote the Lagrangian axes defined in the initial configuration, $\Delta\gamma$ is the discretized plastic strain increment, I_1^σ is the first Cauchy stress invariant and $\tilde{\sigma} = \sqrt{(\sigma' : \sigma')/2}$ is the effective stress, with σ' denoting the Cauchy stress deviator whose spectral decomposition is

$$\sigma' = \sum_A \left(\sigma^A - \frac{I_1^\sigma}{3}\right)\mathbf{N}^A \otimes \mathbf{N}^A. \tag{9.6}$$

The plastic strain increment is given by $\Delta\gamma = \Delta t \, \dot{\gamma}$, where the plastic strain rate is described in this study by the classical Perzyna two-parameter model [196]

$$\dot{\gamma} = \eta\left(\frac{f(\sigma)}{\sigma_Y}\right)^n, \tag{9.7}$$

in which n and η are material constants, σ_Y is the current yield stress and $f(\sigma) = (\tilde{\sigma} - \sigma_Y)$ is the overstress. Strain hardening is captured by introducing a tangent modulus E_t relating the increment of the yield stress, $\Delta\sigma_Y$, to the plastic strain increment, $\Delta\gamma$. Finally, the linear relation

$$\mathbf{S} = \mathbf{LE} \tag{9.8}$$

between the second Piola-Kirchhoff stresses $\mathbf{S}$ and the Lagrangian strains $\mathbf{E}$ is used to describe the elastic response. Assuming material isotropy, the stiffness tensor $\mathbf{L}$ is defined by the Young's modulus E and Poisson's ratio ν.

The main source of load imbalance comes from the very different computational costs associated with the elastic and visco-plastic constitutive updates. As long as the effective stress remains below a given level (chosen in this study as 80% of the yield stress), only the elastic relation (9.8) is computed. Once this threshold is reached for the first time, the visco-plastic update is performed, which typically represents a doubling in the computational cost. Consequently, as the crack propagates through the discretized domain, the load associated with each processor can be substantially heterogeneous due to the plastic zone around the crack tip, thus suggesting the need for a robust dynamic load balancing scheme as described in Section 9.2.

9.3.1   Case Study 1: Elasto-Plastic Wave Propagation

The first application is the quasi-one-dimensional elasto-plastic wave propagation problem depicted in Figure 9.7. It consists of a rectangular bar of length $L = 10$ m and cross-section $A = 1$ m^2. The bar is initially at rest and stress free. It is fixed at one end and subjected at the other end to an applied velocity V ramped linearly from 0 to 20 m/s over .16 ms and then held at a constant velocity of 20 m/s thereafter. The time step size is $3\,\mu s$ and the total number of time steps is 1,100. The material properties are chosen as follows: yield stress $\sigma_Y = 480$ MPa, stiffness $E = 73$ GPa and $E_t = 7.3$ GPa, exponent $n = 0.5$, fluidity $\eta = 10^{-6}$/s, Poisson's ratio $\nu = .33$ and density $\rho = 2800$ kg/m^3.

The applied velocity generates a one-dimensional stress wave that propagates through the bar and reflects from the fixed end. At every wave reflection, the stress level in the bar increases as the end of the bar is continuously pulled at a velocity V. During the initial stage of the dynamic event, the material response is elastic as the first stress wave travels through the bar at the dilatational wave speed $c_d = 6215$ m/s with an amplitude

$$\sigma = \rho c_d V = 348\,\text{MPa} \ < \sigma_Y. \tag{9.9}$$

After one reflection of the wave from the fixed end, the stress level in the bar exceeds the yield stress of the material and the material becomes plastic. A snapshot of the location of the elasto-plastic stress wave is shown in Figure 9.7. The computational overload associated with the plastic update routine (approximately a factor of two increase compared to the elastic case) leads to a significant dynamic load imbalance while the bar transforms from elastic to plastic. As mentioned earlier, in these simulations, the plastic check and update subroutine is called upon when the equivalent stress level exceeds 80% of the yield stress.

The unstructured 800,000-element tetrahedral mesh that spans the bar is initially partitioned into chunks using METIS, and these chunks are then mapped to the processors. During the simulation, the processors advance in lockstep with frequent synchronizing communications required by exchanging of boundary conditions, which may lead to bad performance when load imbalance occurs.

The simulation was run on Tungsten Xeon Linux cluster at the National Center for Supercomputing Applications (NCSA). This cluster is based on Dell PowerEdge 1750 servers, each with two Intel Xeon 3.2 GHz processors, running Red Hat Linux and Myrinet interconnect network. The test ran on

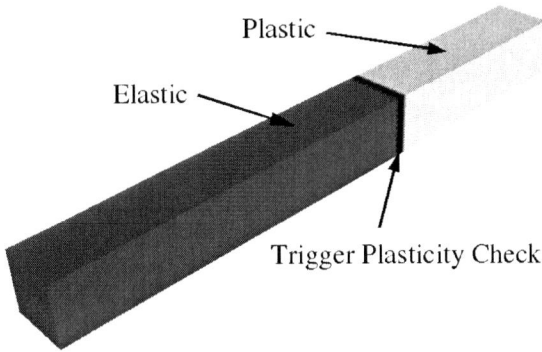

FIGURE 9.7: Location of the traveling elasto-plastic wave at time $c_d t/L = 1.3$.

32 processors with 160 AMPI virtual processors. Figure 9.8 shows the results without load balancing in a CPU utilization graph over a certain time interval. The figure was generated by *Projections* [125], a performance visualization and analysis tool associated with Charm++ that supplies application-level visual and analytical performance feedback. This utilization graph shows how the overall utilization changes as the wave propagates through the bar. The total runtime was 177 seconds for this run.

A separate interest, although not investigated further in this study, is the period of initial load imbalance (observed for the runtime before 48 s in

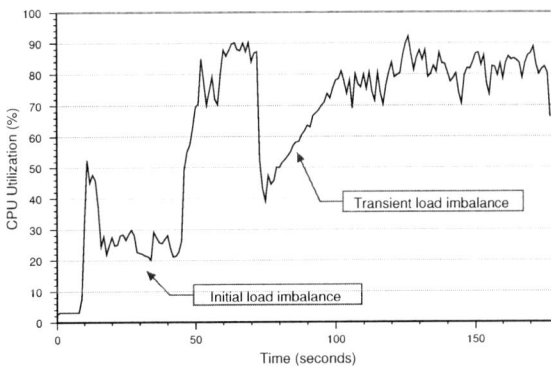

FIGURE 9.8: CPU utilization graph without load balancing (Tungsten Xeon).

Figure 9.8) caused by the quiet generation of subnormal numbers (floating-point numbers that are very close to zero) during the initial propagation of the elastic wave along the initially quiescent bar. This phenomenon is discussed by Lawlor et al. [145], who propose an approach to mitigate such performance effects caused by the inherent processor design. However, this study is only concerned with the load imbalance associated with the transformation of the bar from elastic to plastic (observed for the runtime between 72 s and 100 s in Figure 9.8).

As indicated earlier, the load imbalance in this problem is highly transient, as elements at the wave front change from an elastic to a plastic state. In Figure 9.9, the effects of the plasticity calculations are clearly noticeable in terms of execution time which linearly ramps from the condition of fully elastic to fully plastic resulting in a doubling of the execution time. This leads to a load imbalance, which is resolved by migrating chunks from heavily loaded processors to light ones while the bar goes into the plastic regime.

Even though we used a variety of methods and time frames, the problem was not considerably sped up by load balancing. The transition time was too fast for the load balancer to significantly speed up the simulation. Also the period of imbalance is a very small portion compared to the total runtime. Therefore, a performance improvement here necessitates that the overhead and delays associated with the invocation of the load balancer be minimal. Nevertheless, we managed to speed up the simulation by 7 seconds as shown in

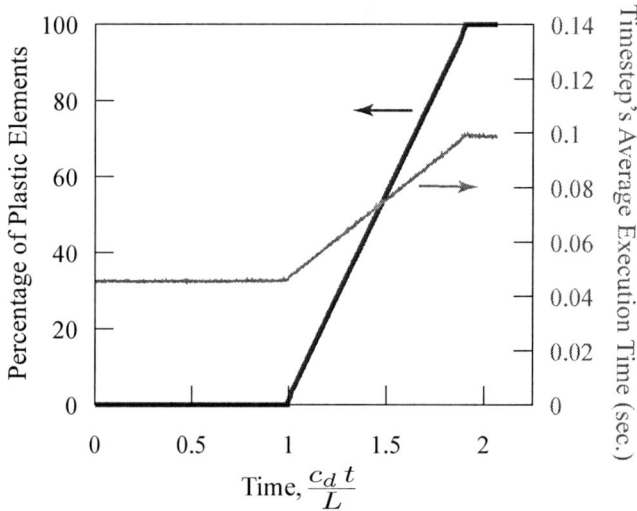

FIGURE 9.9: Evolution of the number of plastic elements.

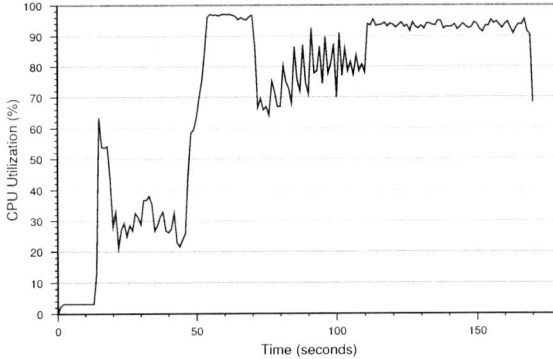

FIGURE 9.10: CPU utilization graph with synchronous load balancing (Tungsten Xeon).

Figure 9.10. The time required for completion reduces to 170 seconds, which yields a 4 percent overall improvement by the load balancing.

We repeated the same test on 32 processors of the SGI Altix (IA64) at NCSA with the same 160 AMPI virtual processors. Figure 9.11 shows the result without load balancing in the Projections utilization graph. The total execution time was 207 seconds and a more severe effect of subnormal numbers on this machine was observed in the first hundred seconds of execution time.

In the second run, we ran the same test with the greedy load balancing scheme described in the previous section. The result is shown in Figure 9.12(a) in the same utilization graph. The load balancing is invoked around time

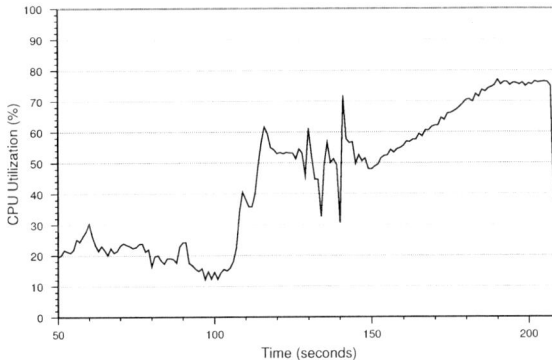

FIGURE 9.11: CPU utilization graph without load balancing (SGI Altix).

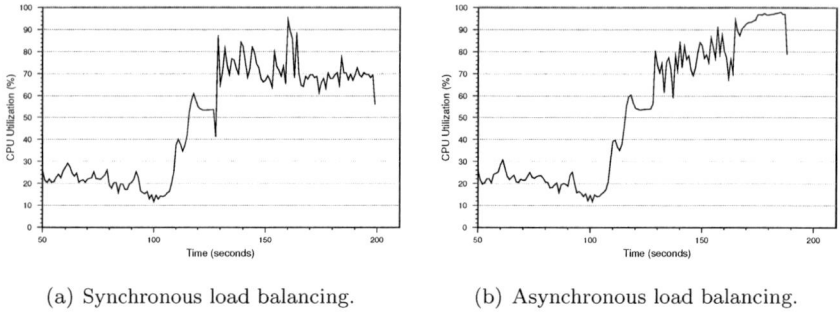

(a) Synchronous load balancing.   (b) Asynchronous load balancing.

FIGURE 9.12: CPU utilization graph (SGI Altix).

interval 130 in the figure. After the load balancing, the CPU utilization is slightly improved and the total execution time is now around 198 seconds.

Finally, we ran the same test with the same greedy algorithm in an asynchronous load balancing scheme described in Section 9.2.5. The asynchronous load balancing scheme avoids the stall of an application for load balancing and overlaps the computation with the load balancing and migration. The result is shown in Figure 9.12(b) in a utilization graph. It can be seen that, after load balancing, the overall CPU utilization was further improved and the total execution time is 187 seconds, which is a 20 second improvement.

9.3.2   Case Study 2: Dynamic Fracture

The second application involves a single edge notched fracture specimen of width $W = 5$ m, height $H = 5$ m, thickness $T = 1$ m and initial crack length $a_0 = 1$ m, having a weakened plane starting at the crack tip and extending along the crack plane to the opposite edge of the specimen. The material properties used in this simulation are $\sigma_Y = 900$ MPa, $E = 210$ GPa, $E_t = 2.4$ GPa, $n = 0.5$, $\eta = 10^{-6}/s$, $\nu = .3$ and $\rho = 7850$ kg/m^3. The boundary conditions along the top and bottom surfaces of the specimen have a linearly ramped velocity of 0.0 to 1.0 m/s over 2.0 ms which is then held at a constant velocity of 1.0 m/s thereafter. The time step size is .47 μs and the total number of time steps is 1.25e5. A single layer of six-node cohesive elements is placed along the weakened interface, with the failure properties described by a critical crack opening displacement value $\Delta_{nc} = .8$ mm and a cohesive failure strength $\sigma_{max} = 95$ MPa. The mesh consists of 91,292 cohesive elements along the interface plane and 4,198,134 linear strain tetrahedral elements. As the stress wave emanating from the top and bottom edges of the specimen reaches the fracture plane, a region of high stress concentration is created around the initial crack tip. In that region, the equivalent stress exceeds the yield stress of the material leading to the creation of a plastic zone. As the stress level

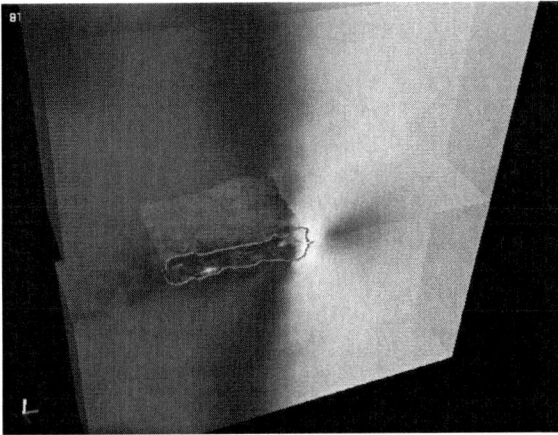

FIGURE 9.13: Snapshot of the plastic zone surrounding the propagating planar crack at time $c_d t / a_0 = 27$. The iso-surfaces denote the extent of the region where the elements have exceeded the yield stress of the material (**see Color Plate 7**).

continues to build up in the vicinity of the crack front, the cohesive tractions along the fracture plane start to exceed the cohesive failure strength of the weakened plane and a crack starts to propagate rapidly along the fracture plane, surrounded by a plastic zone and leaving behind a plastic wake, as illustrated in Figure 9.13.

This simulation was run on the Turing cluster at the University of Illinois at Urbana-Champaign. The cluster consists of 640 dual Apple G5 nodes connected with Myrinet network. The simulation without load balancing took about 12 hours on 100 processors. The evolution of the average processor utilization is shown in the bottom curve of Figure 9.14. As apparent there, around time 10,000 seconds, the CPU utilization dropped from around 85% to only about 44%. This is due to the advent of the elastic elements transitioning into plastic elements around the crack tip, leading to the beginning of load imbalance. As shown in Figure 9.15, the number of plastic elements starts to increase dramatically as the crack starts to propagate along the interface. As more elastic elements turn plastic, the CPU utilization slowly increases and stays around 65% (lower curve in Figure 9.14). The load imbalance can also be easily observed in the CPU utilization graph over processors in Figure 9.16(a). While some of the processors have CPU utilization as high as about 90%, some processors only have about 50% of CPU utilization during the whole execution.

With the greedy load balancing strategy invoked every 500 time steps, the simulation finished in only about 9.5 hours, a saving of nearly 2.5 hours or 20% over the same simulation with no load balancing. This increase is caused by the

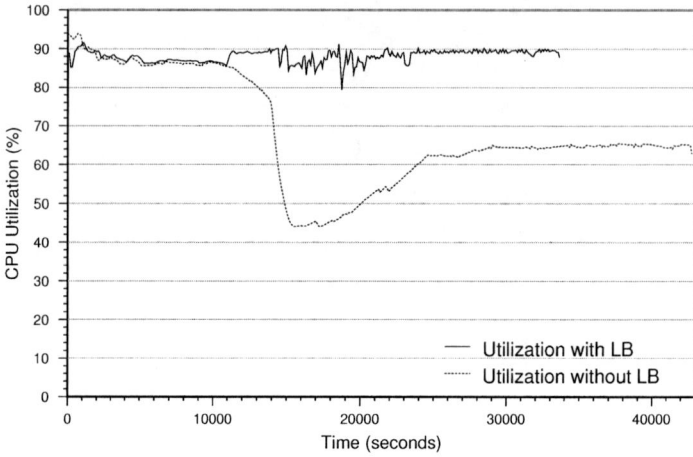

FIGURE 9.14: CPU utilization graph with and without load balancing for the fracture problem shown in Figure 9.13 (Turing Apple Cluster).

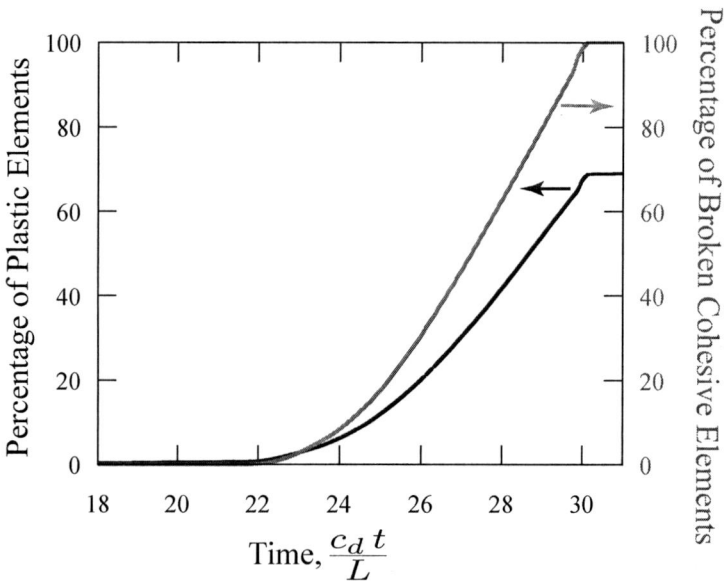

FIGURE 9.15: Evolution of the number of plastic and broken cohesive elements.

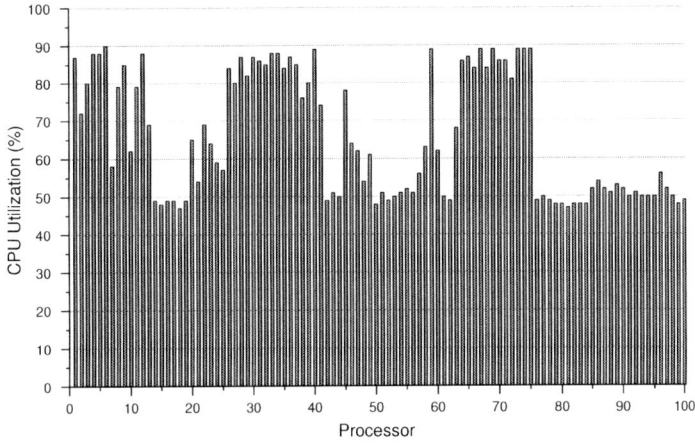

(a) Without load balancing.

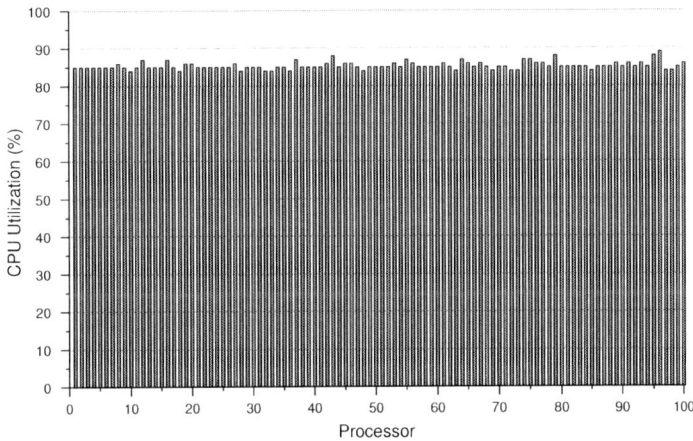

(b) With load balancing.

FIGURE 9.16: CPU utilization across processors (Turing Apple Cluster).

overall increased processor utilization, which can be seen in the upper curve of Figure 9.14. The peaks correspond to the times when the load balancer is activated during the simulation. There is an immediate improvement in the utilization when the load balancer is invoked. Then the performance slowly deteriorates as more elements become plastic. The next invocation tries to balance the load again. Figure 9.16(b) further illustrates that load balance has been improved from Figure 9.16(a) in the view of the CPU utilization across processors. It can be seen that a CPU utilization of around 85% is achieved on all processors with negligible load variance.

9.4   Conclusions

Dynamic and adaptive parallel load balancing is indispensable for handling load imbalance that may arise during a parallel simulation due to mesh adaptation, material nonlinearity and other modern irregular dynamic simulation behavior. We have demonstrated the successful application of the Charm++ measurement-based dynamic load balancing concept to a crack propagation problem, modeled with a cohesive/volumetric finite element scheme. The performance of the application was improved by an agile load balancing strategy which is designed to handle transient load imbalance due to the rapidly propagating wave. The performance study of this application demonstrated the ability of the automatic load balancing to achieve sustained high computational efficiency.

Acknowledgments

This work was an adjunct project affiliated with the Center for Simulation of Advanced Rockets (CSAR) which was funded by the U.S. Department of Energy's Advanced Simulation and Computing program.

Chapter 10

Contagion Diffusion with EpiSimdemics

Keith R. Bisset

Virginia Bioinformatics Institute, Virginia Tech

Ashwin M. Aji, Tariq Kamal, Jae-Seung Yeom and Madhav V. Marathe

Virginia Bioinformatics Institute and Department of Computer Science, Virginia Tech

Eric J. Bohm and Abhishek Gupta

Department of Computer Science, University of Illinois at Urbana-Champaign

10.1 Introduction

Contagion is used here broadly to mean transmitted phenomena such as diseases, opinions, fads, trends, norms, packet diffusion, worm propagation in computer networks, database replication in sensor networks, spread of social movements and influence among peers to purchase music videos or go to movies [88, 165, 152, 94, 56, 36, 178]. The spread of contagions across a national population is a well known complex problem, and includes: (i) pandemics, such as H1N1 and swine influenza outbreaks in recent years, in which the spread of the flu virus is often modeled by stochastic processes, such as the SIR process [185, 59, 73], (ii) spread of information on online social media, such as Twitter and Facebook [207], which are often modeled by stochastic and threshold based models [133] and (iii) the 2003 blackout in Northeastern U.S., and the cascading effects on traffic, communication and other infrastructures that cost $6 billion [174]. A key observation from numerous studies shows that the underlying network structure has a significant impact on the dynamics [84, 221], and as in the case of the 2003 blackout, this could span multiple networks [184].

Though these examples involve very diverse phenomena on different kinds of networks, they often involve very similar and fundamental questions from a dynamical systems perspective. In the case of epidemics, the key challenges include understanding the disease dynamics (e.g., what is the likely attack rate, and who might be vulnerable), detect its onset and peak and develop and evaluate strategies (e.g., vaccinations, school closures) to control the spread of the disease—these have to be translated (often in real time) into policies to be implemented by local and national public health agencies [73, ?]. In the case of the spread of information and viral marketing, it is often of interest to identify the most influential spreaders to initiate the diffusion process, so that the size of the population that gets influenced (to change and adopt a new product, for instance) is maximized [53, 132]. In the case of coupled infrastructures, understanding the impact of cascading failures and identifying criticality of different network elements is a very important issue. Thus, the key questions underlying diverse contagion phenomena can be related to fundamental dynamical systems' problems of control and optimization via exogenous and endogenous resource allocation.

The general form of our problem can be stated informally as follows. We are given one or more networks or interaction graphs. Nodes in the graph represent interactors (computers, sensors, people, agents or any other abstract entity). An edge between two nodes represents an interaction in an abstract form. An example graph is shown in Figure 10.1. Usually the network has spatial structure with a small number of long-range edges and a significant amount of clustering among neighbors. The edges are usually time-dependent. Nodes and edges may have sets of properties that describe physical traits, predispositions

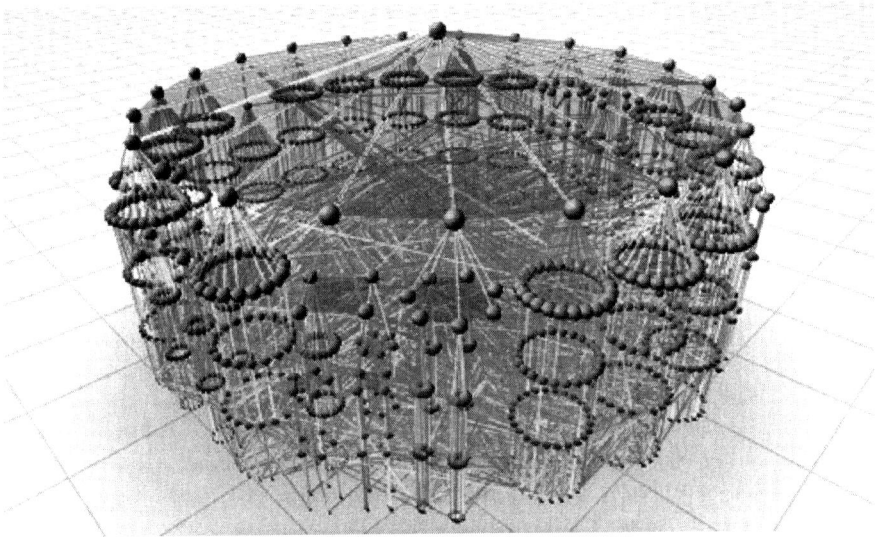

FIGURE 10.1: Example of a synthetic person-person contact network, where people are represented by balls, the size of which is inversely proportional to its degree of separation from the top person. Two people who come in contact during the day are linked by an edge (**see Color Plate 8**). (Image courtesy of David Nadeau, SDSC.)

and beliefs, emotions, strengths of associations and other notions. The problem is to understand dynamic processes on interaction networks, not only baseline dynamics, but also changes in dynamics that result from systematic changes in inputs and external actions. For instance, we can study how failures cascade within and across networks; e.g., if a power line fails, will the subnet go down, or if a substation fails, will it result in a loss of phone connectivity in a certain region?

Simple contagion processes, such as percolation, have been studied extensively and are well understood on simple regular graph families [91]. Analytical closed form results are obtained for many important questions using elegant techniques that include stochastic processes, random graphs and probabilistic methods originating in statistical physics. However, these techniques do not easily extend to large-scale realistic networks, because of the absence of independence and symmetry and the inclusion of heterogeneity. Further, many of the underlying problems related to dynamical properties become computationally intractable (e.g., the probability of a specific individual becoming infected is #P-complete)—thus, high performance computing (HPC) agent-based modeling and simulation (ABMS) techniques thus become a natural choice. Developing computational models to reason about these systems is complicated and scientifically challenging for at least these reasons:

1. Often the size and scale of these systems is extremely large (e.g., pandemic planning at a global scale requires models with 6 billion agents). Further, the networks are highly unstructured and the computations involve complicated dependencies, leading to high communication cost and making standard techniques of load balancing and synchronization ineffective.

2. Individuals are not identical—this implies that models of individual behavioral representation cannot be identical. Behavior depends on individual demographic attributes and the interactions with neighbors [88, 176].

3. The contagion, the underlying interaction network (consisting of both human and technical elements), the public policies and the individual agent behaviors co-evolve making it nearly impossible to apply standard model reduction techniques that are successfully used to study physical systems. For instance, in the case of epidemics, as the disease spreads, people cut down their interactions, thereby sparsifying the network, which in turns slows the disease dynamics.

4. Finally, in many cases as we discuss below, we are faced with modeling multiple networks that are coupled, with possibly multiple contagions evolving in each network.

Additionally, in contrast to many large physics simulations, the outcome of a single run of a socio-technical simulation is not interesting by itself. For example, simulations of infectious disease outbreaks are run not for their own sake, but to investigate specific questions about prevention and mitigation. Answering these questions requires analyzing the interdependent effects of many different parameters, each with many different possible settings, in a stochastic process. The stochastic nature of the simulation, the uncertainty in the initial conditions and the variability of reactions require many replications to develop a bound on the range of results of the simulation. Also, the simulations are not meant to be predictive, but to provide comparative results over a range of simulation inputs. For example, the outcome of an infectious disease propagation experiment is not that a particular number of people will fall ill, but that intervention A is generally more effective than intervention B over the range of expected disease manifestations and initial conditions [73]. Efforts to devise effective policies from simulation experiments are also gaining traction in other domains, such as financial services, e.g., [93].

Typically, overall time-to-solution is the most important measure of simulation effectiveness. Stakeholders need answers as soon as possible. Indeed, many policymakers are forced to use inaccurate, but fast executing, tools in place of accurate but slower tools. Time-to-solution has two components. The first is translation of stakeholder requests, including novel intervention strategies, into a working simulation. For EPISIMDEMICS, this is achieved by designing a simple, flexible domain-specific language for specifying interventions to reduce the amount of simulation code that must be written, and by using Charm++ to more easily bridge the gap between the domain model and the execution model.

The second factor in time-to-solution is the execution of an experimental design. A complete factorial design is often infeasible, and a partial factorial design applied to such a complex, non-linear system can miss important interdependencies. Hence, adaptive designs are usually required. Even with an adaptive design, however, there may still be thousands of experimental cells to explore. The computational burden of a simulation should be considered in the context of such an experiment, and resources should be allocated optimally over hundreds or thousands of runs, each run requiring tens or hundreds of thousands of computational cores.

10.2   Problem Description

An important feature of our work is the formal specification of the problem as a Co-evolving Graphical Discrete Dynamical System (CGDDS) [14, 179]. This formalism provides the ability to reason about the system, and to provide provable guarantees about system performance.

10.2.1   Formalization

Our mathematical model consists of two parts: (i) a discrete dynamical system framework that captures the co-evolution of disease dynamics, social networks and individual behavior and (ii) a **Partially Observable Markov Decision Process** (POMDP) that captures various control and optimization problems formulated on the phase space of this dynamical system.

The discrete dynamical system framework consists of the following components: (i) a collection of entities with state values and local rules for state transitions, (ii) an interaction graph capturing the local dependency of an entity on its neighboring entities and (iii) an update sequence or schedule such that the causality in the system is represented by the composition of local mappings. We can formalize this as follows. A **Co-evolving Graphical Discrete Dynamical System** (CGDDS) S over a given domain $\mathbb{D}$ of state values is a triple $(G(V, E), \mathcal{F}, W)$, whose components are as follows: $G(V, E)$ is the basic underlying social network. V is a set of vertices, and associated with each vertex v_i is a edge modification function g_i. The applications of g_i result in an indexed sequence of social networks. For each vertex v_i, there is a set of local transition functions. The function used to map the state of vertex v_i at time t to its state at time $t + 1$ is f_{v_i, d_t}, and the input to this function is the state sub-configuration induced by the vertex and its neighbors in the contact network $N(i,t)$. The final component is a string W over the alphabet $\{v_1(s), v_2(s), \ldots, v_n(s), v_1(g), \ldots, v_n(g)\}$. The string W is a schedule. It represents an order in which the state of a vertex or the possible edges incident on the vertex will be updated. Both f_i and g_i are assumed probabilistic.

The formalism serves as a mathematical abstraction of the underlying agent-based model. From a modeling perspective, each vertex represents an agent. We assume that the states of the agent come from a finite domain $\mathbb{D}$. The maps $f_{v_i,j}$ are generally stochastic. Computationally, each step of a CGDDS (i.e., the transition from one configuration to another) involves updating either a state associated with a vertex or modifying the set of incident edges on it.

Important Notes. First, we assume that the local transition functions and local graph modification functions are both computable efficiently in polynomial time. Second, the edge modification function as defined can, in one step, simultaneously modify a subset of edges. We have chosen this with the specific application in mind. In all of our applications, when an agent decides to not go to a location (either due to location closure as demanded by public policy or due to the fear of contracting the disease) its edges to all other individuals in that location are simultaneously removed while edges are added to all the individuals who might be at home. Third, the model is assumed to be *Markovian*, in that the updates are based only on the current state of the system, although it is possible to extend the model wherein updates are based on an earlier state of the system. Finally, we assume a parallel update schedule; at even time steps, vertices update their state in parallel and in odd time steps edges of the graph are updated in parallel.

Let $F_{\mathcal{S}}$ denote the **global transition function** associated with $\mathcal{S}$. This function can be viewed either as a function that maps $\mathbb{D}^n$ into $\mathbb{D}^n$ or as a function that maps $\mathbb{D}^V$ into $\mathbb{D}^V$. $F_{\mathcal{S}}$ represents the transitions between configurations, and can therefore be considered as defining the dynamic behavior of a CDDS $\mathcal{S}$. If the local functions or edge update functions are probabilistic then we get a global transition relation with appropriate probabilities on the outgoing configuration nodes. The phase space in this case is simply a large Markov chain $\mathcal{M}$.

In the next step, we overlay a partially observable Markov decision process framework over the discrete dynamical systems framework. This will allow us to discuss control and optimization methods. The discrete dynamical system provides us with a computational view of how state transitions are made. We refer the reader to a recent paper by [180] for detailed definitions and complexity theoretic results on this topic. A POMDP M consists of a finite set of states (S), actions (A) and observations (O). The initial state of the system is $s_o \in S$. t is the local transition function from one state to another and is probabilistic. o is the observation function that assigns to each state an observation that is made. Finally, r is the reward function that tells the reward received when action $a \in A$ when in state s.

Using the terminology of [180] our POMDP is specified succinctly; we use a dynamical systems' specification rather than a circuit representation to achieve this. The states of M are all possible vectors of vertex states and edge states (present or absent). Each vector of the underlying Markov chain $\mathcal{M}$ is specified by a vector of length $\binom{n}{2} + n$, representing all the possible edges and vertices in the graph. The state transitions are obtained by composing the local functions f_i and g_i as discussed above. If these functions are probabilistic then so is the transition function for the Markov chain. Thus the chain consists of $2^{\binom{n}{2}+n}$ states. Actions should be thought of as interventions in our context. Policies map observations to actions. Actions in a state yield reward. The reward (cost) function can be a combination of number of infected individuals and the economic and social costs of the interventions. In this paper, we will primarily concern ourselves with number of infections; the social and economic costs are important and will be discussed in subsequent work. We have two possible classes of reward functions, as is common in game theory: a systemwide reward function and a local reward function associated with each individual. Individuals attempt to maximize their local reward function (e.g., the probability of the individual or a family member becoming infected), while public policy attempts to maximize the systemwide reward function (e.g., the total number of people not infected). The agent-based model described in the next section serves as the corresponding computational model for the POMDP. Partial observability in our context will be captured via various triggering conditions and interventions that are instituted. We will say more about this in the following section.

10.2.2 Application to Computational Epidemiology

We briefly outline how contagion diffusion problems in Computational Epidemiology can be specified using CGDDS. In all the situations considered in this paper, we can make certain simplifying assumptions due to the specific dynamics that we consider. In SIMDEMICS, we have a notion of a **day**. A day is typically 24 hours but can be larger or smaller depending on the specific disease. We *assume* that the social contact network does not dynamically change over the course of a day. By this we mean that at the start of a day, all of the interactions that will occur during the day, including those due to health state changes, public policy interventions and individual behavior, are known. This is a realistic assumption due to the time scale of disease evolution (time it takes for a person to become infectious or symptomatic after being infected). As a result, the schedule can be specified as a sequence of days wherein we only consider disease dynamics over the entire population followed by a step in which there is a change in the social contact network.

Let DS_t denote the one day computation of the disease dynamics over the current social network at time t. Let G denote the social network, A_t denote the change in the activity schedule, while PT_t denotes the change in the within host disease propagation model at time t. Interventions either change the schedule for one or more individuals or change the properties of the within host disease model. Interventions and behavioral changes can be broadly categorized based when they occur as follows.

1. *Non-Adaptive*: Non-adaptive interventions and behavioral change occur before the start of the simulation. The non-adaptive interventions unrealistically assume the population does not change during the course of the epidemic and is limited to studying treatments that have a permanent effect, like vaccination. By a slight abuse of notation, we can represent this as $\mathbf{x}(T) = (D_T \circ \cdots \circ D_0) \circ (A_0 \circ PT_0)(\mathbf{x}(0))$. In other words, we apply our interventions at the start of the simulation and then run the disease propagation model until the end of the simulation.

2. *Adaptive*: The adaptive strategies, on the other hand, incorporate changes in the movement of the people, treatments that have only temporary effects (antiviral medications are only effective when being taken) and wholesale changes to the interactions within the population (like school closure). This is represented most generally as $\mathbf{x}(t) = F^t(\mathbf{x}(0))$ as $(D \circ PT \circ A)^t(\mathbf{x}(0))$. Here $\mathbf{x}(t)$ denotes the system state at time t. We can now differentiate various strategies by how frequently PT and A are applied as compared to D. In other words, we view the dynamics as the following composition: $(D^{t/r} \circ PT \circ A)^r(\mathbf{x}(0))$, where the exponents reflect the different time scales. This can be viewed as degree of adaptation. Policy based change in the social network is usually caused by changing the behavior of a set of individuals in some uniform way. Furthermore, it is natural to expect that these changes do not occur often. In contrast, individual behavior based changes can occur every day — individuals can change their behavior and thus their probability of contracting a disease on a daily basis. A simulation is computationally most efficient when t is small, since it amounts to fewer updates to the social network and individual behavior, while making t small makes the simulation less realistic since the interaction between individual behavior and disease dynamics is not well represented.

10.3   EpiSimdemics Design

The EPISIMDEMICS model is used to explore the impact of agent behavior and public policy mitigation strategies on the spread of contagion over extremely large interaction networks. To determine their local behavior, individual agents may consider their individual state (the contagion model), their demographics and the global state of the simulation. Contagion and behavior are modeled as coupled **Probabilistic Timed Transition Systems (PTTS)**, an extension of finite state machines with two additional features:

the state transitions are probabilistic and timed. A PTTS is a set of states. Each state has an id, a set of attributes' values (the same attributes for each state of the PTTS), a dwell time distribution and one or more labeled sets of weighted transitions to other states. The label on the transition sets is used to select the appropriate set of transitions, given the pharmaceutical treatments that have been applied to an individual. The attributes of a state describe the levels of infectivity, susceptibility and symptoms an individual who is in that state possess. Once an individual enters a state, the amount of time that they will remain in that state is drawn from the dwell time distribution.

Between-host contagion transmission and within-host contagion progression can be viewed as two connected but independent processes. Between-host transmission triggers the start of within-host progression by causing an uninfected individual to transition to an infected state. The disease progress of the infected individual is then fully determined by the local function governing the within-host progression. The within-host disease progression is modeled as a PTTS. The system also supports multiple interacting PTTSs for modeling of multiple co-circulating diseases, enhanced sociological modeling in the agents and the addition of more complex interventions, such as contact tracing and antiviral stockpiles.

The PTTS and the interaction network are co-evolving, as the progression of each one potentially affects the other. In simple terms, who you meet determines if you fall sick, and the progression of the disease may change who you meet (e.g., you stay home because you are sick). The co-evolution can be much more complex, as an individual's schedule may change depending on information exchanged with others, the health state of people they contact even if no infection takes place (e.g., more people than usual are sick at work) or even expected contacts that do not happen (e.g., coworkers who are absent from work). All of this may also be affected by an individual's demographics (e.g., a person's income affects their decision to stay home from work). The interaction network can represent different types of interactions such as physical proximity or telephone communication. The discussion here is restricted to computational epidemiology of infectious diseases such as influenza, but the model itself is quite general. It has been recently used to study the spread of malware in a wireless communication network [37], and the interaction of cells within the human immune system [24].

The interaction network used in this work is a highly resolved model of individuals and their movement throughout the day. Briefly, it is the result of a complex data fusion process that merges information from sources such as the United States Census, NAVTEQ Street data, the National Household Travel Survey from the US Department of Transportation, information about business locations from Dun & Bradstreet's Strategic Database Marketing Records, information about primary and secondary education facilities from the US Department of Education and several others. This process is mathematically rigorous and produces a statistically valid synthetic population. A more detailed description of this process can be found in [19, 11, 183, 15].

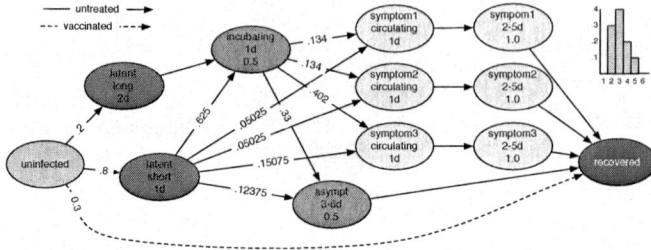

FIGURE 10.2: PTTS for the H5N1 disease model. Ovals represent disease state, while lines represent the transition between states, labeled with the transition probabilities. The line type represents the treatment applied to an individual. The states contain a label and the dwell time within the state, and the infectivity if other than 1.

10.3.1 The Disease Model

The disease propagation (inter-host) and disease progression (intra-host) models were developed in the National Institutes of Health Models of Infectious Disease Agent Study (MIDAS) project [181]. The disease progression model is shown in Figure 10.2.

When a susceptible individual and an infectious individual are colocated, the propagation of disease from the infected individual to the susceptible individual is modeled by

$$p_{i \to j} = 1 - (1 - r_i s_j \rho)^\tau \tag{10.1}$$

where $p_{i \to j}$ is the probability of infectious individual i infecting susceptible individual j, τ is the duration of exposure, r_i is the infectivity of i, s_j is the susceptibility of j and ρ is the transmissibility, a disease-specific property defined as the probability of a single completely susceptible person being infected by a single completely infectious person during one minute of exposure [12]. Generally, ρ is calibrated to produce a desired attack rate (fraction of total population infected) in the absence of any interventions.

10.3.2 Modeling Behavior of Individual Agents

Each individual agent has a set of possible activity lists, called schedules, for different purposes. Examples include a normative schedule, one for school closures, one to use while staying home when sick, etc. Each schedule has an associated type. Each agent also keeps a priority list of schedule types. The current schedule is created from the schedule type with the highest priority. There can only be one schedule type for a given priority level which can be changed or removed at any time by the scenario. This is used to manage the currently active schedule for an agent. For example, take the case of a worker who stays home to care for his children when schools are closed, and then falls sick. When he is recovered, if schools are still closed he remains at home, but if schools have reopened he will return to work. His normative schedule has priority 1. When schools are closed, he is given a school closure schedule of priority 2. When he falls sick, he is given a stay at home schedule of priority 3. If schools are reopened while he is sick, his priority 2 schedule is removed, but he continues to follow his priority 3 schedule. When he recovers, his priority 3 schedule is removed, and he switches to the remaining schedule with the highest priority. Currently, all of the schedules must be precomputed. Ongoing work will add the capability to generate schedules on demand. Some schedules, such as self-isolation at home, can be dynamically created as needed to save memory and decrease startup time. Others, such as going to the nearest medical facility, depend on an individual's location when the schedule change is invoked, and need to be computed dynamically as the simulation runs.

10.3.3 Intervention and Behavior Modification

The scenario specifies the behavior of individuals (e.g., stay home when sick), as well as public policy (e.g., school closure when a specific proportion of the students are sick). There are two fundamental changes that can be made that will effect the spread of a contagion in a social network. All behavior and public policy interventions can be implemented through these changes. First, the probability of transmission of a contagion can be changed by changing the infectivity or susceptibility of one or more individuals. For example, getting vaccinated reduces an individual's susceptibility, while wearing a mask while sick reduces an individual's infectivity. Taking antiviral medication, such as TamiFlu (oseltamivir), reduces the likelihood of becoming infected and reduces both the infectivity and length of the infectious period once an infection has taken place. Second, edges can be added, removed or altered in the social network, resulting in different individuals coming into contact for different amounts of time. The individual behaviors and public policy interventions in EPISIMDEMICS, collectively referred to as the scenario, expose these two changes in a way that is flexible, easy to understand for the modeler and computationally efficient.

The scenario is a series of triggers and actions written in a domain-specific language. The grammar for this language is shown in Figure 10.3. While conceptually simple, this language has proven to be quite powerful in describing a large range of interventions and public policies. A trigger is a conditional statement that is applied to each individual separately. If a trigger evaluates to true, one or more actions are executed. These actions can modify the individual by changing its attributes or schedule type, explicitly transitioning the PTTS and modifying scenario variables. Scenario variables can be written (assigned, incremented and decremented) and read in the scenario file. The value read is always the value at the end of the previous simulation day. Any writes to a scenario variable are accumulated locally and synchronized among processors at the end of each simulated day.

⟨scenario⟩ → **version** ⟨maojr⟩.⟨minor⟩
        (⟨intervention⟩ | ⟨trigger⟩ | ⟨comment⟩)*;
⟨intervention⟩ → **intervention** ⟨intervention_name⟩ ⟨action⟩+;
⟨trigger⟩ → **trigger** [**repeatable**] [**single**] [**with prob** = ⟨real⟩]
⟨condition⟩ ⟨action⟩ |
        state (**on entry** | **on exit**) ⟨state_name⟩ [**with prob** = ⟨real⟩]
⟨condition⟩ ⟨action⟩;
⟨action⟩ →
        apply ⟨intervention_name⟩ [**with prob**= (⟨real⟩ | ⟨real_var⟩)] |
        treat ⟨fsm_name⟩ ⟨treatment_name⟩ |
        untreat ⟨fsm_name⟩ ⟨treatment_name⟩ |
        schedule ⟨sched_name⟩ ⟨priority⟩ |
        unschedule ⟨priority⟩ |
        infect ⟨fsm_name⟩ |
        transition ⟨fsm_name⟩[:⟨state_name⟩] [**keeptime** | normal] |
        create ⟨integer⟩ ⟨integer⟩ [**apply** ⟨intervention_name⟩] |
        remove |
        endsim |
        message ⟨string⟩ |
        set (⟨var_name⟩ | **person.**⟨person_attribute⟩) (= ⟨integer⟩ | **++**
| -- | += ⟨integer⟩ | -= ⟨integer⟩) |
        add (⟨int_var⟩ | ⟨integer⟩ **to** ⟨set_name⟩ |
        delete (⟨int_var⟩ | ⟨integer⟩ **from** ⟨set_name⟩);
⟨condition⟩ → ⟨or_expr⟩;
⟨or_expr⟩ → ⟨and_expr⟩ | **or** ⟨and_expr⟩;
⟨and_expr⟩ → ⟨not_expt⟩ | **and** ⟨not_expr⟩;
⟨not_expt⟩ → **not** ⟨or_expr⟩ | **(** ⟨or_expr⟩ **)** | ⟨base_expr⟩;
⟨base_expr⟩ → ⟨binary_cond⟩ | ⟨set_cond⟩ | **true** | **false**;
⟨binary_cond⟩ →
        ⟨int_var⟩ ⟨binary_op⟩ ⟨integer⟩ |
        ⟨real_var⟩ ⟨binary_op⟩ ⟨real⟩ |
        ⟨string_var⟩ ⟨binary_op⟩ ⟨string⟩;
⟨set_cond⟩ → ⟨set_name⟩ **intersect** ⟨set_name⟩ **is not null** |
        ⟨set_name⟩ **contains** (⟨int_var⟩ | ⟨integer⟩));
⟨binary_op⟩ → < | <= | = | != | >= | >;
⟨var⟩ → ⟨int_var⟩ | ⟨real_var⟩ | ⟨string_var⟩;
⟨int_var⟩ →
        day |
        time |
        person.id |
        person.removed |
        person.⟨person_attribute⟩ |
        ⟨fsm_name⟩**.infected** |
        ⟨fsm_name⟩**.**⟨fsm_attribute⟩ |
        ⟨var_name⟩);
⟨real_var⟩ → ⟨fsm_name⟩**.infectivity** | ⟨fsm_name⟩**.susceptibility**;
⟨string_var⟩ → ⟨fsm_name⟩**.state**;
⟨*XXX*_name⟩ → [**a-zA-Z0-9_**]+;
⟨string⟩ → "[**a-zA-Z0-9_**]+";
⟨comment⟩ → **#** .* ⟨EOL⟩;

FIGURE 10.3: Grammar of EPISIMDEMICS scenario file.

When the PTTS is manually transitioned, the new state can either be explicitly specified or chosen as part of the normal transition process (i.e., the new state is selected from the weighted edges that are part of the transition set with the correct label). There are also several ways of determining the dwell time in the new state: (i) Pick the dwell time from the distribution in the new state, (ii) Keep the dwell time from the old state, (iii) Pick the dwell time from the distribution in the new state, and subtract the amount of time already spent in the old state, performing another transition according to the normal transition rules if this results in a dwell time that is equal to or less than zero and (iv) Pick the dwell time from the distribution in the new state, but reduce it by the percentage of the dwell time spent in the old state.

10.3.4   Social Network Representation

In EPISIMDEMICS, the social network is represented by a labeled bipartite graph, Figure 10.4(a), where nodes consist of people and locations, referred to as a people-location graph. If a person visits a location, there is an edge between them, labeled by the type of activity and the time of the visit. Each node (person and location) can also have labels. The labels attached to persons correspond to his/her demographic attributes such as age or income. The labels attached to locations specify the location's attributes such as its geographic coordinates, the types of activity performed there, maximum capacity, etc. It is important to note that there can be multiple edges between a person and a location which record different visits. Internally, this network is converted into the equivalent person-person graph, as shown in Figure 10.4(c). This graph has an edge between two people if they were in the same location at the same time, labeled by the time of contact. This form of the contact network is much more continent for calculating interactions between people, but much less sparse, containing approximately 10 times more edges than the person-location graph. Figure 10.4(b) shows the people that are colocated in space and time. Assuming that person 2 is infected and either in the latent state (infectious contagious, but not yet symptomatic) or infectious (contagious and symptomatic), Figure 10.4(d) shows the potential transmission. The social contact graph is not static, but changes over time in response to changes in a person's health state (e.g., stay home when sick), public policy (e.g., school closure) or behavior changes (e.g., reduce unnecessary activities during an outbreak). This new network, in turn, affects a person's health (e.g., reducing contact with potentially infectious individuals outside the home, or increasing contact with potentially infected children inside the home). Including this co-evolution is important for correctly modeling the spread of disease [13].

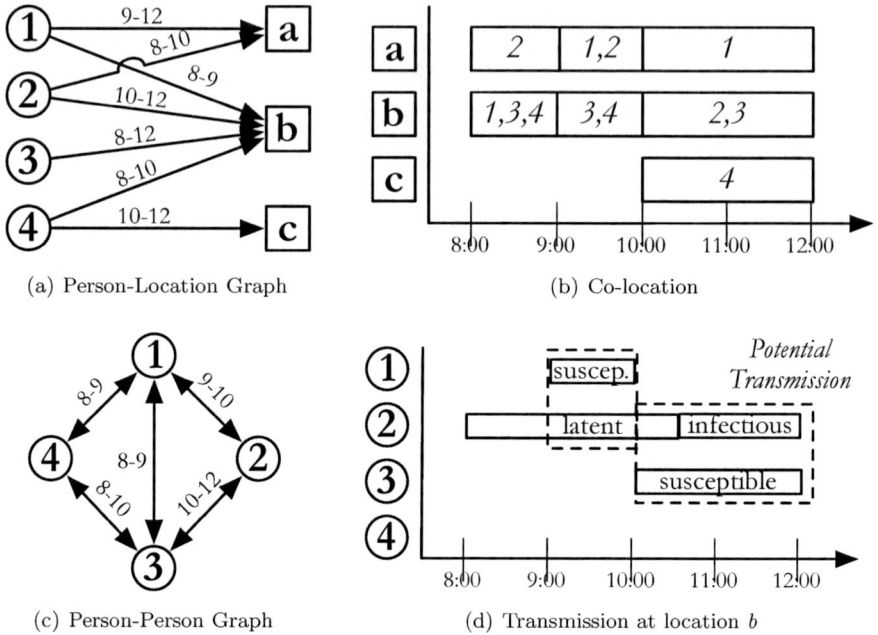

(a) Person-Location Graph

(b) Co-location

(c) Person-Person Graph

(d) Transmission at location b

FIGURE 10.4: An example social contact network: (a) the bipartite graph representation showing people visiting locations; (b) the temporal and spatial projection of the network; (c) the person-person graph showing interactions between temporally and spatially co-located people; (d) potential disease transmission between an infectious person *2*, and susceptible people *1* and *3*.

10.4   EpiSimdemics Algorithm

The EPISIMDEMICS model can be simulated with a simple discrete event simulation (DES) algorithm in which the system only changes its state upon the occurrence of an event. As shown in Figure 10.4(d), there are two types of events in the system: *Arrive Events* (person p arrives at location l at time t_{arrive}) and *Depart Events* (person p leaves location l at time t_{depart}).

To ensure correctness, the algorithm has to adhere to the following causality constraint: *If an individual i leaves location L_A at time t_{depart} and arrives at location L_B at time t_{arrive}, his/her health state when arriving at L_B (denoted by $s_i(t_{arrive})$) has to be decided prior to calculating the states of other individuals at L_B after time t_B.* This causality constraint leads to temporal and spatial dependencies among nodes in the simulated system.

For simplicity of exposition, travel between locations is shown as instantaneous. In the actual system, there is a delay between leaving one location and arriving at the next location, based on an approximation of the travel time between locations. This delay can be calculated with varying degrees of accuracy [11]. Typically, disease transmission on public transportation vehicles has been ignored, but can be easily modeled by adding buses, trains, etc., as additional locations.

By sorting all events by a global clock and updating the health states of all individuals (according to the disease model) as well as their next destination locations (according to their original schedule and dynamic responses to the environment), we can obtain an implementation for simulating a coupled PTTS. However, for a large population consisting of hundreds of millions of individuals, the computational resources provided by a single machine are insufficient.

There are three important semantic points of the contagion diffusion problem that lead to the EPISIMDEMICS algorithm.

1. Individuals can only affect other individuals through interactions that occur when they are co-located in space and time.

2. An individual's health state changes, once infected, can be precomputed.

3. There is a minimum latent period, D_{min}. This is the amount of time that must pass between a person becoming infected and a person being able to infect others. For most infectious diseases, there is a suitable latent period that is determined by the biology of the infection. For influenza, this period is at least 24 hours.

The above observations led to a semantics-oriented problem decomposition. The existence of a latent period for newly infected individuals in the disease model provides a basis for relaxing the global clock. If the time period to be simulated is divided into n iterations, and if the *length of single simulation iteration* is less than D_{min}, then all locations can be concurrently considered and interactions between individuals at these locations can be simulated in parallel. This is the basis of the EPISIMDEMICS algorithm, illustrated in Algorithm 3.

Based on discussions from the previous section, we organize the operations of the sequential EPISIMDEMICS algorithm during each simulation phase into the structure shown in Figure 10.5. This structure depicts the three types of computational components: persons, locations and the communication subsystem, as well as concurrencies in their local computation and communications. In this structure, the communication subsystem mediates the communication between processes.

The processing is separated into iterations that represent, for influenza, simulated days. It is important to note that state changes are not limited to time step boundaries. For example, if an individual is infected at 10:47 on day 10, and becomes infectious 36 hours later, he can start infecting others at 22:47 on day 11. Each iteration has the basic following steps:

Input: A social network $G = (P, L, V)$, a disease manifestation M, an initial system state $S(0)$, a set of scenarios C, a simulation iteration Δt and a total simulation time T

Output: A sequence of system states over time $S(\Delta t), S(2\Delta t), \ldots, S(T)$

1 **for** $t = 0$ **to** T *increasing by* Δt **do**

    foreach *individual* $i \in P$ **do**

        compute a set of visits $v_{i,j} = (l_{i,j}, t_{i,j}, d_{i,j}, s_{i,j}), 1 \leq j \leq N_i$, where N_i is the number of locations i will visit during the time interval $(t, t + \Delta t)$, and $l_{i,j}$, $t_{i,j}$, $d_{i,j}$ and $s_{i,j}$ denote the target location, the start time, the duration and i's health state during its jth visit;

        foreach $v_{i,j}$ **do**

            if $s_{i,j}$ *changes value to* $s'_{i,j}$ *at time* t' *and* $t_{i,j} < t' < t_{i,j} + d_{i,j}$ **then**

                split $v_{i,j}$ into two visits $v_{i,j}^a = (l_{i,j}, t_{i,j}, d_{i,j}^a, s_{i,j})$ and $v_{i,j}^b = (l_{i,j}, t', d_{i,j}^b, s'_{i,j})$, where $d_{i,j}^a = t' - t_{i,j}$ and $d_{i,j}^b = d_{i,j} - d_{i,j}^a$

            end

        end

        foreach $v_{i,j}$ **do**

            send $v_{i,j}$ to location l_j;

        end

    end

2   **foreach** *location* $l_j \in L$ **do**

        compose a serial DES;

        foreach *visit* $v_{i,j}$ **do**

            add an arrive event $e_{i,j}^A$ at time $t_{i,j}$, and a depart event $e_{i,j}^D$ at time $t_{i,j} + d_{i,j}$ to a priority queue Q_j in which events are ordered by processing time

        end

        foreach *event* $e \in Q_j$ **do**

            remove e from the head of Q_{L_j};

            compute the outcome $r_{i,j}$ of e;

            send $r_{i,j}$ to person p_i

        end

    end

3   **foreach** *individual* $i \in P$ **do**

        receive and combine all results $r_{i,j}$;

        update i's health state

    end

end

4 update global simulation state;

Algorithm 3: The EPISIMDEMICS algorithm.

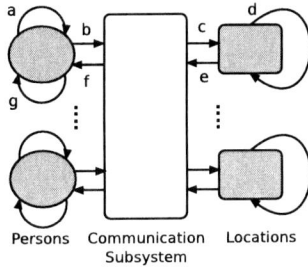

	Sequence of operations
a:	person compute visits
b:	person send visits
c:	location receive visits
d:	location compute visits
e:	location send outcomes
f:	person receive outcomes
g:	person combine outcomes

FIGURE 10.5: The computational structure of the sequential EPISIM-DEMICS algorithm.

1. Each individual determines the locations that they are going to visit, based on a normative schedule, public policy and individual behavior and health state. The person sends a message to each visited location with the details of the visit (time, duration and health state during the visit). This can be computed in parallel for each person.

2. Each location computes the interactions that occur between occupants of that location. Each interaction may or may not result in an infection, depending on a stochastic model. For the Epidemiological model, disease propagation is modeled by Equation 10.1.

   A message is then sent to each infected person with the details of the infection (time of infection, infector and location). Again, each location can perform this computation in parallel, once it has received information for all of the individuals that will visit the location during the iteration.

3. Each person who was infected updates his health state by transitioning out of the infected state. In the event that the person was infected in multiple locations, the earliest infection is chosen.

4. Any global simulation state (i.e., scenario variables) is updated.

For each iteration, there are two synchronizations required: between steps 1 and 2 and between steps 2 and 3. In addition, step 4 requires a reduction operation.

10.5   Charm++ Implementation

This section describes different entities in EPISIMDEMICS and the features of Charm++ that are used to develop EPISIMDEMICS. Next follows the EPISIMDEMICS algorithm and details of interaction of different entities in EPISIMDEMICS. We also highlight the importance of the Charm++ platform for our algorithm in terms of programmer's productivity and performance.

10.5.1   Designing the Chares

The EPISIMDEMICS algorithm involves iterative message exchanges between the set of person objects and the set of location objects, for the desired number of simulation days. In the MPI implementation, we distributed the person and location objects in a round-robin fashion among the available MPI processes, and message passing occurred by explicitly calling the MPI communication primitives. A PersonManager and LocationManager object on each process manage the communication for the associated object. Since the number and destination of messages is unknown until they are actually generated, each Manager sends an empty message to every other Manager of the same type to signal that all of the messages have been sent. In the Charm++ version, as shown in Figure 10.6, we create two types of *chare arrays* called LocationManager and PersonManager to contain the manager chares. The Charm++ runtime maps the chares to processes and runs the simulation. In summary, we follow a two-level hierarchical data distribution technique—first we assign people/locations to chares; next we let the Charm++ runtime assign chares to processes. We later show that an efficient assignment of chares to processes can result in a very efficient communication pattern amongst the simulation objects. We also show that the load balancing strategies used in our application gives us a better load distribution compared to not using any load balancing.

At the start of the simulation we initialize routines for modules such as loggers and file parsers. We create a *chare group* called InitManager for this purpose, where a single chare group object is instantiated in every process,

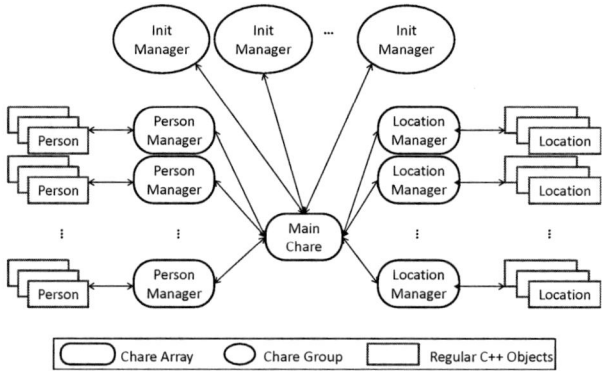

FIGURE 10.6: Entities in EPISIMDEMICS.

as shown in Figure 10.6. The per-process initializations are invoked at the beginning of the simulation in every chare group object, so that all the other chares of the system can access the required handles and other resources from their local chare group objects.

The stochastic nature of EPISIMDEMICS *ideally* requires a unique random stream object for each location and person object in the system to be perfectly repeatable. Repeatability is desirable for several reasons: it eases the debugging burden, it ensures that data can be regenerated in the future as long as the configuration files are kept and it allows branching while keeping the initial portions of the simulation identical. It would require a significant amount of memory to allocate a separate stream for hundreds of millions of objects. To alleviate this problem, the MPI version of EPISIMDEMICS creates one random stream per process, so that the location and person objects within each process can share random streams. While we can run repeatable simulation when using the same number of MPI processes, the output may change when changing the processor count. Similar limitations apply to EPISIMDEMICS as well, because we share random number streams within a Chare array element object, which contains multiple people or location objects. With EPISIMDEMICS, this repeatability is maintained even when a Charm++ object migrates to a different processor. It is a trade-off between the flexibility of simulation and efficiency of memory management. If we maintain the total number of chares in the system a constant, we can potentially have the same number of total random number streams for any number of processors. If the number of processors in the system is changed, the Charm++ runtime will simply create a different allocation of chares to processors, but the total chare count (and the total random number streams) remains constant. Thus, we reduce memory overhead yet we can produce repeatable simulations.

One of the useful features of Charm++ is dynamic load balancing via chare migration. Charm++ allows us to specify the phase in the program during which load balancing should occur, and also lets us choose the specific chare arrays that have to be migrated. In EPISIMDEMICS, we migrate PersonManager and LocationManager chares at the end of some pre-determined iterations (simulation days). The LocationManager chares compute the interactions for a simulated day and do not store any state information at the end of each iteration. LocationManager chare elements are ideal for migration as they are lightweight and the computation varies dynamically as people change their schedule due to illness or public policy interventions such as school closure. Our current implementation uses only measured computational load, using the built-in load balancers of Charm++ to decide which chares to migrate. We plan to use the facilities of Charm++ to use simulation semantics, including knowledge of future changes to load (e.g., upcoming interventions) to increase load balancing effectiveness.

10.5.2   The EpiSimdemics Algorithm

In this section we talk about some of the important entities of Charm++ used in our application. We also show the interaction among these entities. The Main chare first creates the LocationManager and PersonManager chare arrays. It then creates the InitManager chare group objects to do per-process initializations and later to maintain the global simulation state. The proxies (or handles) to all the chare array objects are defined as global *read-only* objects, so that any chare can invoke the entry methods on every other chare in the system. Next, the input files (locations, persons and visits) are read by the Main chare and distributed randomly to the appropriate LocationManager or PersonManager chare objects. Once all the input is read in to memory, the simulation process is ready to start.

The different stages of a single iteration in the EPISIMDEMICS algorithm are shown in Figure 10.7. At the beginning of each simulation day, the person objects compute the visit messages and send them to the location objects, i.e., the PersonManager chare elements send messages to specific LocationManager chare elements (Figure 10.7(a)). The location objects compute the infections and send messages to the person objects about the new infections for the current simulation day (Figure 10.7(c)). The person objects then decide the locations that have to be visited on the next iteration and the process continues. The Charm++ runtime system implicitly handles queuing, dispatch and processing of messages at the source and destination chares/processes. Moreover all our messages are sent asynchronously, thereby enabling efficient overlap of communication with computation.

The EPISIMDEMICS algorithm has two global synchronization points per iteration. The first is to ensure that every visit message sent from a person has been received at the destination location before the location starts its computations. The second is to ensure that every interaction result message sent by a location has been received by the destination person before the person's state is updated. EPISIMDEMICS synchronization is a combination of the contribute() and QD() features of Charm++. The contribute() (global reduction) makes sure that all the necessary events are produced. The Quiescence Detection (QD) makes sure that all the produced events are consumed. At the first synchronization point, each of the PersonManager chare calls *contribute()* when it is done with producing all its events (all of its associated person objects have sent all of their visit messages). This is followed by a QD call that will finish when there are no messages in flight across the entire system. This ensures that every visit message that is produced has been consumed on the LocationManager side and it is safe to proceed to the next step. The second global synchronization is analogous except that here the producer is LocationManager and the consumer is PersonManager. It is important to note that both the global reduction and the QD can be asynchronous and additional work can be done during the synchronization, as long as no additional

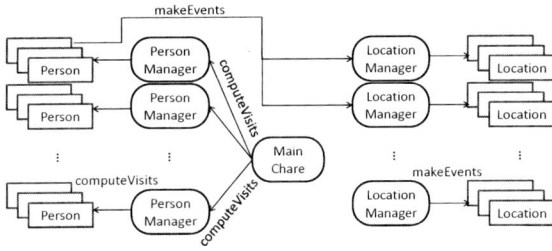

(a) Step 1: Computation and transfer of the visit messages.

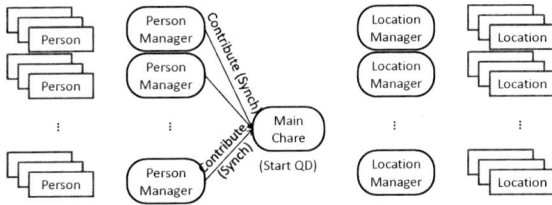

(b) Step 2 (Barrier-1): Synchronize with the main chare; wait via *Quiescence Detection* for visit messages to be delivered.

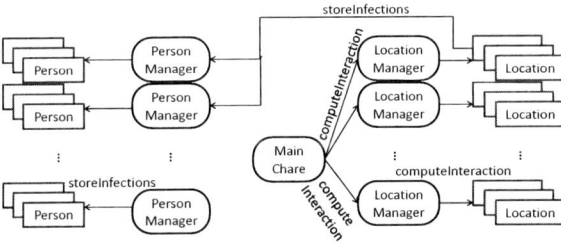

(c) Step 3: Computation and transfer of the infection messages.

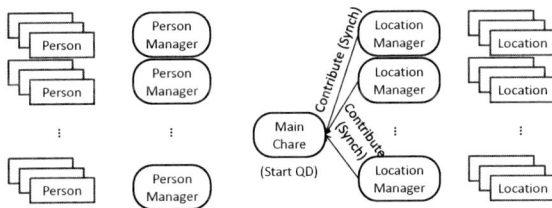

(d) Step 4 (Barrier-2): Synchronize with the main chare; wait via *Quiescence Detection* for infection messages to be delivered.

FIGURE 10.7: The EPISIMDEMICS Algorithm: One Simulation Day.

message traffic is generated during the QD. We show in Section 10.6 that the time taken for QD is less than 5% of the total execution time.

At the end of each iteration, the PersonManager chare array elements resolve all the received infection messages and updates the person state. At the same time, the LocationManager chare elements can be allowed to migrate for load balancing purposes, because they do not have any more computations to perform for the simulation day. The Charm++ runtime can potentially perform the above two actions simultaneously, thereby improving performance. Charm++ helps improve the productivity of the programmer significantly, because the object oriented nature of Charm++ allows the programmer to just focus on the implementations of the individual chare array elements and leaves the communication optimizations to the framework.

10.5.3   Charm++ Features

EPISIMDEMICS has driven the research and development of several Charm++ features to get better performance and productivity. Here we summarize some of them.

In order to generate efficient replicas as the simulation is progressing, EPISIMDEMICS needs a way to create a copy of the existing chare arrays. To support this functionality, we created a chare array copy mechanism such that the application needs to call a function arrayCopy and the rest is handled by the runtime system. Since this array creation and copy happens in the middle of execution, the application developer needs to make sure that it happens at a synchronization point. After the copy is complete, control is passed to the application via a user-supplied callback function.

Another challenge when performing multiple simulations using replicas is the fact that quiescence detection is a global phenomenon, while the synchronization is only needed within individual array copies. To circumvent this, we use a lightweight Completion Detection (CD) mechanism which is local to a module and hence other copies can keep making progress while quiescence is being detected on a subset of objects in the system.

EPISIMDEMICS uses Charm++'s pupDisk mechanism for efficient disk read and data distribution on application startup. Using this framework, the overhead of reading data from disk is distributed across multiple nodes compared to the scheme where data is read sequentially on a master node and distributed to remaining nodes through interprocess communication.

10.6   Performance of EpiSimdemics

In this section, we first identify the compute-intensive methods in EPISIMDEMICS. Then, we evaluate the strong and weak scaling performance. Next, we evaluate the impact of applying load balancing on the performance of EPISIMDEMICS. We also show the efficacy of the quiescence detection method for synchronization in EPISIMDEMICS.

10.6.1 Experimental Setup

Hardware: We use two hardware platforms for our experiments. For the experiments with processor utilization, synchronization, load balancing and scenario intervention, respectively, discussed in Section 10.6.2, 10.6.3, 10.6.6 and 10.7, we use a cluster consisting of 96 compute nodes. Each node has two quad-core Intel Xeon E5440 processors and 16 GB of DDR2 memory. The nodes are connected by 20 Gb/s InfiniBand interconnects. For the scaling experiments discussed in Section 10.6.4 and 10.6.5, we use the NSF's Blue Waters at NCSA. Blue Waters is a Cray XE6/XK7 installation at the NCSA NPCF. Though it contains both XE and XK nodes, our experiments were confined to the 22,640 XE nodes. Each XE node has 2 AMD 6276 Interlagos processors, one Gemini NIC connected in a 3D Torus, and 64 GB of memory, with 102.4 GB/s of memory bandwidth. Each Interlagos processor is comprised of 8 Bulldozer core modules, with 2 integer units and 1 floating point unit per module, resulting in 16 core modules per node, 362,240 for the entire XE machine.

Software: We use Charm++ v6.5 built with gemini_gni-craxyxe for the scaling experiments discussed in Section 10.6.4 and 10.6.5. We use Charm++ v6.2 built with net-linux-x86_64 support for the load balancing experiments discussed in Section 10.6.6, and v6.3 built with net-linux-x86_64 support for the rest of our experiments.

Data: EPISIMDEMICS can simulate the spread of the H5N1 avian influenza virus across social contact networks of various populations. Typical sizes of input population range between 0.5 million and 280 million. The population data is organized by US states. We choose a single state or a group of states to discuss different aspects of our experiments. Each test is executed for 120 iterations, where each iteration corresponds to a simulation day. We configure our simulation such that the attack rate, the fraction of the total population ultimately infected, in the absence of any interventions, is roughly 55%.

10.6.2 Performance Characteristics

We first investigate which of the EPISIMDEMICS methods dominate the overall execution time. Our simulation data for this test was the population of Virginia (6.8 million people), and we used 40 processor cores for execution. We rely on the Charm++ framework to collect detailed execution time statistics, and take advantage of the Charm++ performance visualization tool, *Projections* [127]. For this experiment, we enabled load balancing, but will defer its analysis and discussion to Section 10.6.6.

Figure 10.8 shows the breakdown of the CPU utilization among the se-

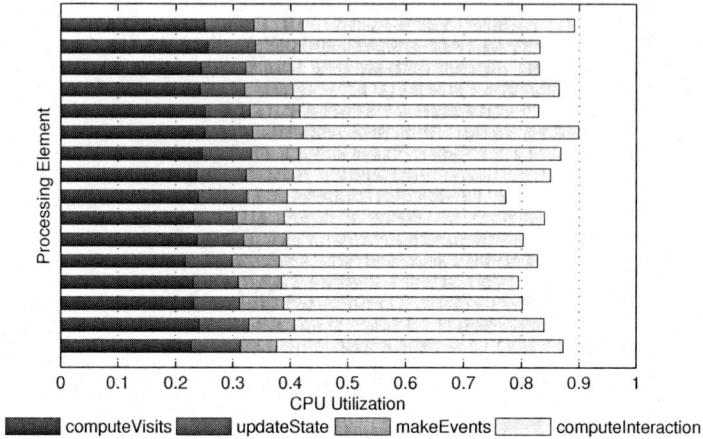

FIGURE 10.8: The breakdown of CPU utilization identifies compute-intensive Charm++ methods. On average, computeInteraction() takes 46% of the execution time, computeVisits() takes 22%, updateState() takes 8.2% and makeEvents() takes 7.8% of the execution time.

lected EPISIMDEMICS methods from the entire simulation run, excluding initialization and input data reading. Due to space limitations, we plot the results only for a subset of the processors used. LocationManager's computeInteraction() method takes 46% of the total execution time, and it is the most time consuming. The computeVisits() method determines each person's daily schedule and sends visit messages to LocationManager chares constituting 22% of the total execution time. PersonManager's updateState() method takes 8.2% of the time for updating the person's status at the end of each simulation day. LocationManager's makeEvents() adds the messages received from PersonManager chares to local data structures and takes 7.8% of the execution time.

10.6.3   Effects of Synchronization

Figure 10.9 shows the synchronization cost relative to the total execution time. As discussed in Section 10.5, EPISIMDEMICS relies on contribute() and QD for synchronization. This method takes at most 4.23% of the total execution time in our experiments. We employ QD for our experiments except for the scaling ones where we use CD instead. It is difficult to measure the cost of synchronization using CD because each producer object locally initiates the synchronization process as soon as it is done with producing events.

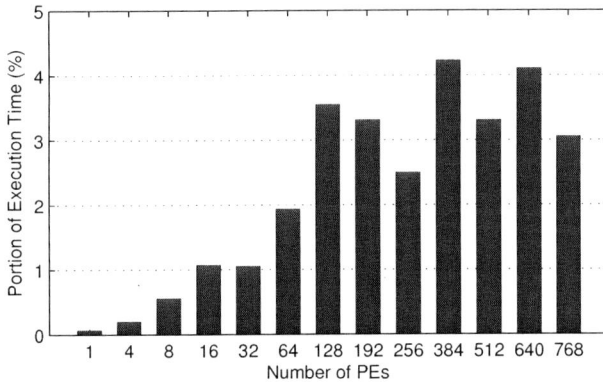

FIGURE 10.9: The synchronization cost using contribute() and QD method takes at most 4.23% of the total execution time for Arkansas data using up to 768 PEs.

10.6.4 Effects of Strong Scaling

In this section we compare the scalability of EPISIMDEMICS with a fixed problem size, as reported in [248]. We use the entire US data of 280 million people. Figures 10.10 shows the total execution time and the speedup. We extrapolate the runtime for a single core module based on the observed runtime using 240 core-modules as the basis for calculating speedup. For this experiment, load balancing is not enabled and we assign one chare of each

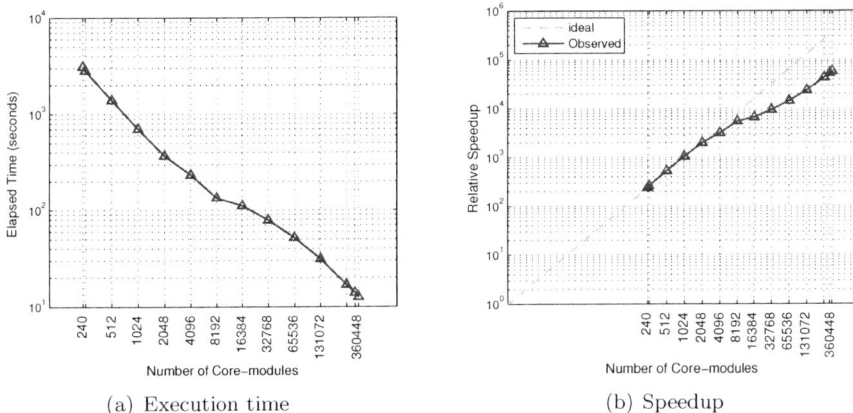

(a) Execution time

(b) Speedup

FIGURE 10.10: Scalability of EPISIMDEMICS with a fixed problem size. EPISIMDEMICS achieves up to 58,649× speedup for the entire US data using 352K core-modules which is 16.3% of the ideal speedup.

type per processor. EPISIMDEMICS scales up to 352K core-modules and gains a speedup of up to 58,649 which is 16.3% of the ideal speedup.

10.6.5   Effects of Weak Scaling

To show the scalability of our implementation for large social contact network data, we increase the input data size as we increase the number of compute nodes such that the population per node remains constant (which we denote as POP_{target}). We choose to use $POP_{target} = 2,190,607$ which is roughly close to 280,397,680/128. For a given number of nodes, we pick a certain set of states such that the total population from the chosen states is close to $POP_{target} \times N_{nodes}$, where N_{nodes} is the number of nodes used. Since the size of the data from the different states is not likely a multiple of POP_{target}, the population per node varies within 0.41% as we pick a different number of nodes and different set of data to run. Thus, for comparing the performance with data of different sizes, we normalize an execution time T_{org} taken to run POP_{actual} sized data by calculating the representative execution time as $T_{rep} = T_{org} \times POP_{target}/POP_{actual}$. Further details of datasets are listed in Table 10.1. It lists the total population of data, POP_{actual}, T_{org}, T_{rep}, as well as the list of states of which data are chosen.

Figure 10.11 shows the weak scaling performance as reported in [248]. When we use 128 nodes (2048 core-modules), the problem size is 128 times larger than when we use one node (16 core-modules). As the per-node population is the same, ideally, the execution time should be constant. We consider the ideal time T_{ideal} as the execution time of EPISIMDEMICS for the data of POP_{target} population using a single node. Since the Oklahoma data and the New Mexico data together contain roughly the same population as POP_{target}, we calculate T_{ideal} to be 340 seconds by the normalization as $T_{\{ID,SD\}} \times POP_{target}/POP_{\{ID,SD\}}$, where $T_{\{ID,SD\}}$ denotes the execution time to run EPISIMDEMICS for the Idaho data and the South Dakota data

Nodes / Core-modules	People ($\times 10^6$)	Visits ($\times 10^6$)	Locations ($\times 10^6$)	Execution Time (s) measured	normalized
1/16	2.2	12.1	0.57	341	340
2/32	4.4	24.1	1.20	348	349
4/64	8.8	48.3	2.50	354	353
8/128	17.5	96.4	4.42	352	354
16/256	35	193	8.57	356	356
32/512	70	385	16.0	358	358
64/1024	140	770	34.2	359	360
128/2048	280	1541	71.7	366	366

TABLE 10.1: Data chosen for the weak scaling experiment shown in Figure 10.11. As we increase the number of nodes, we use the data of larger population such that each node simulates roughly 2,190,607 people.

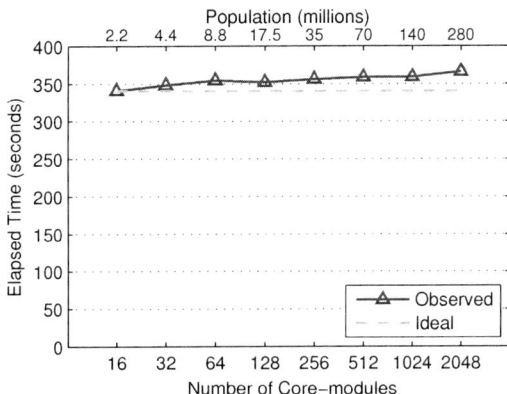

FIGURE 10.11: Scalability of EpiSimdemics with a constant problem size per compute node. EpiSimdemics executes 8% longer than the ideal using 128 nodes (2048 core-modules).

together using a single node (16 core-modules). EpiSimdemics executes 8% longer than T_{ideal} using 128 nodes (2048 core-modules).

10.6.6   Effect of Load Balancing

The Charm++ framework migrates compute-intensive tasks (chare objects) from heavily loaded processors to less utilized processors. Charm++ load balancing can be done in a centralized, fully distributed or hierarchical fashion. With the centralized approach, the machine's load and communication structure are all accumulated to one processor, followed by a decision process that determines the new distribution of Charm++ objects at given synchronization points. We are using a centralized approach in our simulation, as it best suits to applications that iterate over multiple phases. We use the *Greedy* and *Refine* centralized load balancing strategies in our experiments. GreedyLB uses a greedy algorithm to assign chares to processors. It does not attempt to minimize the amount of chare migration that occurs. On the other hand, RefineLB limits the number of chares migrated by shifting chares from highly utilized processors to underutilized processors to reach average. As our first step of load migration in our application, we are migrating LocationManager chares only, as they are lightweight and easy to migrate.

There should be several chares on one processor for the migration to happen from one processor to another. At the same time, some overhead is incurred with increasing the number of chares on a processor. Figure 10.12 shows the increase in execution time for increased number of chares. For our load balancing experiments we use two LocationManager chares per processor.

Figure 10.13 shows the effect of load balancing strategies applied when simulating the Arkansas population data. This diagram is generated using

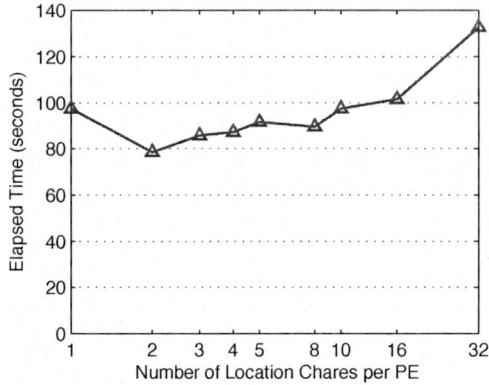

FIGURE 10.12: Impact of the number of chares per PE used on load balancing. For simulating the North Carolina population, the execution takes 98 seconds using total 128 PEs and a single chare per PE (without load balancing). With two chares per PE, it takes 78 seconds which is about 81% of that without load balancing.

the Charm++ Projection tool [127]. The horizontal bars show PEs, while the colors show methods and idle times. Since there is a large overhead with GreedyLB, we apply it only once after the first iteration and follow it up by a RefineLB step after every third iteration. The dark orange color shows the time taken by computeInteraction() and the white following it shows the idle time. Three out of twelve processors in the first iteration have extended idle times (white lines) which mean that the computational loads of computeInteraction() are not even and other PEs are waiting on them to finish their compute step. At the beginning of the second iteration, GreedyLB (shown by red color) is applied, which aggressively migrates objects from heavily loaded processors to less utilized processors. In the second iteration the load is more balanced

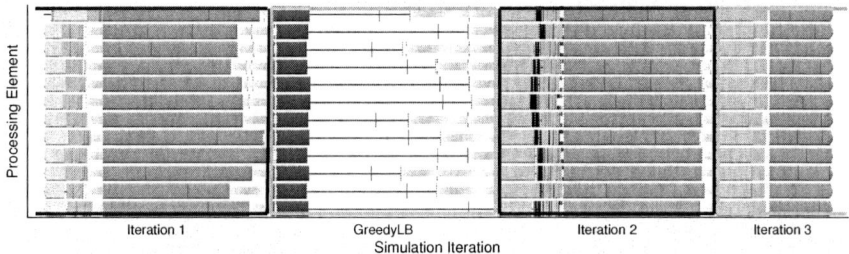

FIGURE 10.13: Effect of Load Balancing: heavily imbalanced in Iteration-1 (three PEs have more load than the others). GreedyLB at the start of second iteration reallocates chares to processes with lighter loads. Iteration-2 shows better load distribution.

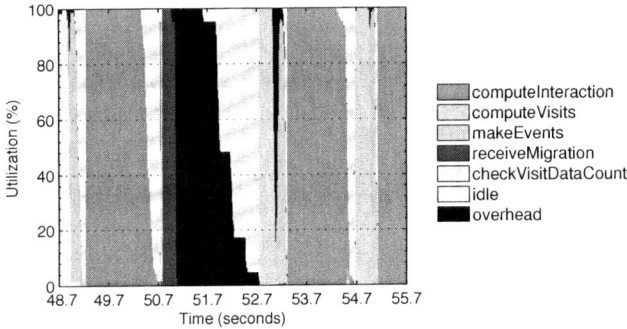

FIGURE 10.14: Waterfall graph shows CPU utilization across all processes by methods. Red shows the load migration while the black shows the overhead from load migration plus the unaccounted time used by the Charm++ framework. The white areas show processor idle times.

compared to the first iteration (small wait times after computeInteraction). We further apply RefineLB after every third iteration to reduce load imbalance through the iterations.

Figure 10.14 shows the CPU utilization by EPISIMDEMICS methods, the load balancing overhead and idle times over two simulation days. Note that the communication of computeVisits() overlaps the computation of makeEvents() methods. The remote entry method makeEvents() called within computeVisits() to send visit messages from PersonManager to remote LocationManager is a non-blocking call and put into the runtime queue until the CPU resource becomes available on the remote processors. There are also less compute-intensive methods, but because of their smaller proportion to total utilization they are not visible. We can see the overhead from GreedyLB step (shown by red color) after the first iteration. We can also notice in the iteration after the GreedyLB is applied that the computeInteraction() load is more balanced across all the processors. The black color shows the overhead associated with GreedyLB and the unaccounted time taken by the Charm++ framework. The gaps (white in color) between utilization sections show the idling during synchronizations. Figure 10.9 shows the time taken by Quiescence Detection (synchronization), running Arkansas data with one chare of each type per process. The QD time is maximum (4.3%) when running on 348 processors. This shows that QD is not taking a lot of time even when running a large number of processors.

To get a good view of the benefits of load balancing and show the individual contribution of Greedy and Refine steps, we show time per iteration for simulation running over 60 iteration days. For this experiment we used GreedyLB and a combination of GreedyLB and RefineLB. Figure 10.15 shows the time/iteration performance of load balancing strategies compared to not

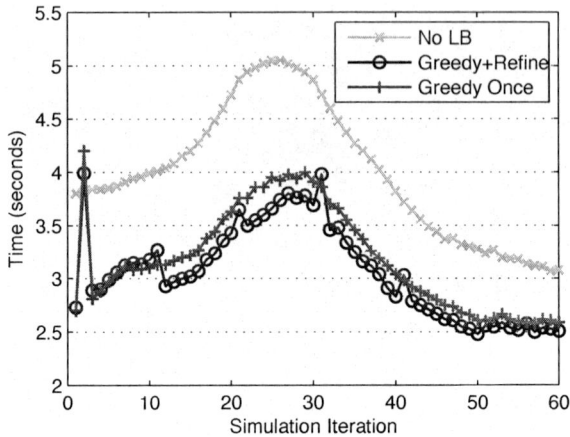

FIGURE 10.15: Performance gain for different load balancing strategies for NC data on 128 PEs. Total simulation time when not using load balancing is 240 seconds, GreedyLB applied only once after the 1st iteration is 191 seconds and GreedyLB followed by refine steps every 10th iteration is 181 seconds.

using it. All the results in this figure are collected for North Carolina data (6.8 million population) on 128 processors with two chares per process. Load balancing is performed for LocationManager chares only, which are the most compute-intensive tasks in our application (take about 46% of total compute time).

"No LB" means that no load balancing is happening. "Greedy once" means GreedyLB is applied once after the first iteration. "Greedy+Refine" means that GreedyLB is applied after the first iteration followed by RefineLB every 10th iteration. The longer times for the second iteration show the overhead associated with GreedyLB step. Notice the reduction in time per iteration when GreedyLB is applied at the second iteration. In case of GreedyLB, load balancing happens only once in the entire simulation. However, for Greedy+Refine, the GreedyLB in the second iteration is followed by refine steps every tenth iteration. This can be seen at the start of the 11th iteration when there is visible drop in time per iteration after that. The same refine step effect can be seen in iterations 21, 31, 41 and 51. Cumulative timings show that GreedyLB when used only once after the first iteration gives about 20 percent performance improvement compared to not using the load balancing (No LB). Similarly, the performance improvement of GreedyLB+RefineLB scheme compared to not using load balancing is more than 24 percent.

10.7   Representative Study

We describe an example of a case study performed at the request of one of our sponsors. While the example is slightly modified from the actual study in order to simplify it, and to present a more complete representation of the features present in EPISIMDEMICS, it provides an accurate representation of multiple studies that have been done. Due to space constraints, it omits details that may be important to answer specific questions. In a pandemic, a well-matched vaccine may not be immediately available and only become available in batches. These vaccines should be distributed according to some prioritization scheme to ensure maximum effectiveness. Other studies investigating the optimal vaccination problem are described in [83, 235].

In the study presented here, the interactions between the health care system and infected individuals are ignored. This is an important topic and is an area of active research. One study [154] modeled the delays in health care seeking, where the presence of symptoms and day of week were factored into the likelihood of an individual seeking care. This, in turn, affected which cases would show up in the surveillance system, impacting the policy maker's view of the epidemic. Another study [161] looked at the factors involved in making health decisions, specifically seeking medicine and paying for it, given local and global knowledge of the unfolding epidemic. We have also begun to study Nosocomial transmission of disease, i.e., the transmission of disease within the health care system [111].

The various disease parameters used below (e.g., transmission rates, incubation period, compliance rates) are reasonable values based on previous work and collaboration with medical and public health experts. In order to determine the efficacy levels of vaccines and other pharmaceuticals we use the latest published literature. In some cases, such as early in the 2009 H1N1 pandemic, exact characteristics of the disease are just not known. In such cases a sensitivity analysis is performed to provide information about the effect that these values have on the outcome, and are refined as more information is obtained from the field.

The behavioral aspects of the model are more difficult to calibrate. Most of them don't have literature to back them up at the detail level that is required. Depending on context, either reasonable behaviors are modeled based on expert opinion, or a range of behaviors is modeled [161, 162]. One advantage of detailed modeling systems like EPISIMDEMICS is that these kinds of questions can be explored in great detail, posing a variety of "What if" questions.

The population of the six east coast states, CT, MD, MA, NH, NJ and NY (41.6 million residents, 9.9 million locations and 208.8 million activities of six types), is simulated. The contagion that will be spread is the H5N1 influenza virus (i.e., Avian flu). Figure 10.2 shows the PTTS associated with the H5N1 disease model.

There are four different interventions that may be applied to different groups at different times: vaccination (V), self-isolation (I), school closure (S) and anti-viral treatment (A). School closure only affects the population of children, whose age is less than 18, while the other intervention affects three different age groups: children, adults ($18 \leq$ age < 65) and seniors (age $\geq$ 65). The vaccine is assumed to be low efficacy, meaning that only 30% of the individuals vaccinated will be protected. This is representative of a vaccine for a newly emerging strain of influenza. When schools are closed, an adult care giver is required to remain home with any children under 13 years of age. Sixty percent of the school-children are considered to be compliant, meaning that they remain at home for their entire day. The other 40% participate in their normal after-school activities. Critical workers will be isolated in some type of group quarters in small groups of 16, where they do not come in contact with people outside of their subgroup. When an individual chooses to self-isolate, they withdraw to home and do not participate in any activities away from home. They still contact other people in their household during the times they are at home together. Table 10.2 describes the four sets of interventions and Figure 10.16 shows part of the scenario definition.

The intra-host disease progression passes through several stages [191, 94]. When a person is exposed to the disease, it starts off in a latent stage, where the individual is neither infectious or symptomatic, potentially followed by an incubating stage during which the individual is partially infectious and still does not exhibit symptoms. In the final stage of the disease an individual is fully infectious and displays one of four levels of symptoms from asymptomatic to fully symptomatic. The probability of a person staying home instead of participating in their normal activities increases with the severity of their symptoms. Once the disease has run its course, the infected individual is considered recovered and can not be reinfected.

Vaccination takes effect at the start of the simulation; the other interventions are triggered when a certain percentage of a subpopulation is diagnosed with the virus. It is assumed that 60% of those who are symptomatic will be diagnosed. School closure is done on a county by county basis, based on the number of sick children who reside in each county. Each person who enters one of the *symptomX* states has a probability of withdrawing to home, depending on the severity of the symptoms (*symptom1*—20%, *symptom2*—50%, *symptom3*—95%).

Sixteen scenarios were simulated, specifying all combinations of the four sets of interventions (I, Q, S, V). Figure 10.17 shows the number of individuals that are infected for each cell. They can be grouped into four categories: those without vaccination or school closure (labeled **a** in the figure), vaccination (labeled **b**), school closure (labeled **c**) and both vaccinations and school closure (labeled **d**). Self-isolation and quarantine of critical workers have little impact on the overall infection rates. Self-isolation happens late into the epidemic (2.5% of the adult population is infected on a single day), limiting its effectiveness. In fact, when school closure or vaccination is included, the trig-

Label	Intervention	Compliance	Trigger	Description
V	Vaccinate Adults Vaccinate Children Vaccinate Critical Workers	25% 60% 100%	Day0	Prevent 30% of treated individuals from becoming infected.
S	School Closure School Reopen	60%	1.0% Children diagnosed (by county) 0.1% Children diagnosed (by county)	Children stay home during school hours (with adult if under 13). Compliant children stay home outside of school hours.
Q	Quarantine Critical Workers	100%	1.0% adult population diagnosed	Critical workers removed from general population and isolated in small groups.
I	Self-isolation	20%	2.5% adult population diagnosed	Eliminate all activities outside of the home.

TABLE 10.2: Description of the experimental setup showing the sets of interventions, their compliance rates and when the intervention takes effect.

```
 1 trigger disease.infected = 1
 2   set cur_infected++
 3
 4 trigger disease.state = "disease:Incapacitated:removed"
 5   set cur_infected--
 6
 7 trigger disease.infected = 1 and person.age < 18
 8   set cur_infected_child++
 9
10 trigger disease.state = "disease:Incapacitated:removed"
11         and person.age < 18
12   set cur_infected_child--
13
14 # Diagnosed by health care professional
15 intervention diagnose
16     set person.diagnosed = 1
17     set total_diagnosed++
18
19 trigger disease.symptom > 1
20     apply diagnose with prob=0.75
21
22 ###### vaccination intervention
23 intervention vaccinate
24     treat disease vaccine
25     set vaccinated++
26
27 # vaccination rates from
28 # http://www.cdc.gov/mmwr/preview/mmwrhtml/su6001a7.htm
29 trigger person.age < 18
30     apply vaccinate with prob=0.55
31
32 ###### selfisolation intervention
33 intervention selfisolate
34     schedule stayAtHome 1
35     set selfisolated++
36
37 # Symptom level one is non-specific symptoms
38 trigger person.age < 18 and  disease.symptom = 1
39     apply selfisolate with prob= 0.15
40
41 # symptom level 2
42 # selfisolation rates from BMC Public Health.
43 trigger person.age < 18 and  disease.symptom = 2
44     apply selfisolate with prob= 0.75
45
46 # Symptom level three requires medical attention
47 trigger disease.symptom >= 3
48     apply selfisolate with prob=1.0
```

FIGURE 10.16: Part of a scenario file which defines four types of interventions. This scenario file has 2630 non-empty lines and applies to 129 different counties from six east coast states.

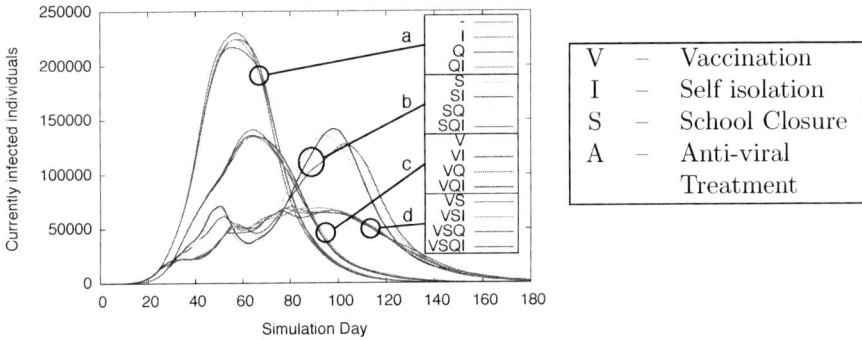

V	–	Vaccination
I	–	Self isolation
S	–	School Closure
A	–	Anti-viral
		Treatment

FIGURE 10.17: Currently infected individuals by day for all 16 combinations of 4 interventions.

ger level is never reached. Quarantine of critical workers affects such a small portion of the total population (about a third of a percent) that its effects are not apparent in the total population. However, quarantine reduces the percentage of critical workers who are infected from 40% without quarantine to 18% when critical workers are both quarantined and vaccinated, which may be vitally important for maintaining a functioning society.

Another interesting phenomenon happens with school closure. Schools are closed when 1% of school children are diagnosed as ill on a particular day in a county. The schools are reopened, on a county by county basis, when the number of diagnosed children falls below 0.1%. Even at that low level, there is enough residual infection to cause another wave of infections. This can be observed in the dual peaks of the group of epicurves labeled **b** in Figure 10.17. There is also a significant geographic component to the spread of infection. Figure 10.18 shows the time period for each county during which the schools are closed, ordered by state and then the start of closure, and the percentage of children infected for each county in Maryland. One of the most important advantage of individual level models is the ability to do this type of detailed analysis to determine the *why* behind the results.

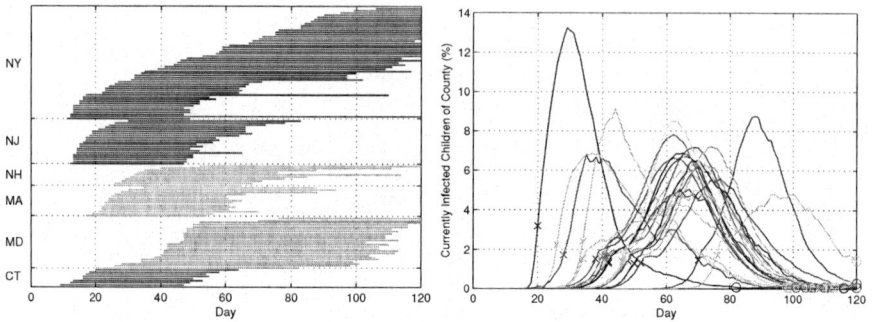

FIGURE 10.18: The school closure timing by county, sorted by initial closure day within each state (left); the percentage of infected children in each county of MD when school closure intervention is applied. The day when the school closure begins effective for each county is shown by ×, and reopening is shown by ○ (right) **(see Color Plate 9)**.

Acknowledgments

The quality EPISIMDEMICS software and the models behind them are due to many individuals in addition to the authors. We would like to thank the past and present members of NDSSL and our collaborators. This work has been partially supported by NSF PetaApps Grant OCI-0904844, NSF PRAC Grant OCI-0832603, NSF NetSE Grant CNS-1011769, DTRA Grant HDTRA1-11-1-0016, DTRA CNIMS Contract HDTRA1-11-D-0016-0001 and the LLNL George Michael Fellowship.

This research is part of the Blue Waters sustained-petascale computing project, which is supported by NSF award number OCI 07-25070 and the state of Illinois. Blue Waters is a joint effort of Illinois and NCSA. This work is also part of the Contagion PRAC allocation support by NSF award number OCI-0832603.

Bibliography

[1] A. Adcroft, C. Hill, and J. Marshall. Representation of topography by shaved cells in a height coordinate ocean model. *Monthly Weather Review*, 125(9):2293–2315, 1997.

[2] L. Adhianto, S. Banerjee, M. Fagan, M. Krentel, G. Marin, J. Mellor-Crummey, and N. R. Tallent. Hpctoolkit: Tools for performance analysis of optimized parallel programs http://hpctoolkit.org. *Concurr. Comput.: Pract. Exper.*, 22:685–701, April 2010.

[3] M.P. Allen and D.J. Tildesley. *Computer Simulations of Liquids*. Claredon Press, Oxford, (1989).

[4] R. J. Anderson. Tree data structures for n-body simulation. *SIAM J. Comput.*, 28:1923–1940, 1999.

[5] R. E. Angulo, V. Springel, S. D. M. White, A. Jenkins, C. M. Baugh, and C. S. Frenk. Scaling relations for galaxy clusters in the Millennium-XXL simulation. *ArXiv e-prints*, March 2012.

[6] Gabriel Antoniu, Luc Bouge, and Raymond Namyst. An efficient and transparent thread migration scheme in the PM^2 runtime system. In *Proc. 3rd Workshop on Runtime Systems for Parallel Programming (RTSPP) San Juan, Puerto Rico. Lecture Notes in Computer Science 1586*, pages 496–510. Springer-Verlag, April 1999.

[7] Amnon Barak, Shai Guday, and Richard G. Wheeler. The mosix distributed operating system. In *LNCS 672*. Springer, 1993.

[8] Kevin Barker, Andrey Chernikov, Nikos Chrisochoides, and Keshav Pingali. A Load Balancing Framework for Adaptive and Asynchronous Applications. In *IEEE Transactions on Parallel and Distributed Systems*, volume 15, pages 183–192, 2003.

[9] Kevin J. Barker and Nikos P. Chrisochoides. An Evaluation of a Framework for the Dynamic Load Balancing of Highly Adaptive and Irregular Parallel Applications. In *Proceedings of SC 2003*, Phoenix, AZ, 2003.

[10] J. Barnes and P. Hut. A Hierarchical O(NlogN) Force-Calculation Algorithm. *Nature*, 324:446–449, December 1986.

[11] C. Barrett, R. Beckman, K. Berkbigler, K. Bisset, B. Bush, K. Campbell, S. Eubank, K. Henson, J. Hurford, D. Kubicek, M. Marathe, P. Romero, J. Smith, L. Smith, P. Speckman, P. Stretz, G. Thayer, E. Eeckhout, and M. Williams. TRANSIMS: Transportation Analysis Simulation System. Technical Report LA-UR-00-1725, LANL, 2001.

[12] C. L. Barrett, K. Bisset, S. Eubank, M. V. Marathe, V.S. Anil Kumar, and Henning Mortveit. *Modeling and Simulation of Biological Networks*, chapter Modeling and Simulation of Large Biological, Information and Socio-Technical Systems: An Interaction Based Approach, pages 101–147. AMS, 2007.

[13] C. L. Barrett, S. Eubank, and M. V. Marathe. An interaction based approach to computational epidemics. In *AAAI' 08: Proceedings of the Annual Conference of AAAI*, Chicago USA, 2008. AAAI Press.

[14] C. L. Barrett, H. B. Hunt III, M. V. Marathe, S. S. Ravi, D. J. Rosenkrantz, and R. E. Stearns. Complexity of Reachability Problems for Finite Discrete Dynamical Systems. *J. Comput. Syst. Sci.*, 72(8):1317–1345, 2006.

[15] Christopher L. Barrett, Richard J. Beckman, Maleq Khan, V.S. Anil Kumar, Madhav V. Marathe, Paula E. Stretz, Tridib Dutta, and Bryan Lewis. Generation and analysis of large synthetic social contact networks. In M. D. Rossetti, R. R. Hill, B. Johansson, A. Dunkin, and R. G. Ingalls, editors, *Proceedings of the 2009 Winter Simulation Conference*, Piscataway, New Jersey, December 2009. Institute of Electrical and Electronics Engineers, Inc.

[16] A. Basermann, J. Clinckemaillie, T. Coupez, J. Fingberg, H. Digonnet, R. Ducloux, J.-M. Gratien, U. Hartmann, G. Lonsdale, B. Maerten, D. Roose, and C. Walshaw. Dynamic load balancing of finite element applications with the DRAMA Library. In *Applied Math. Modeling*, volume 25, pages 83–98, 2000.

[17] Jerome Baudry, Emad Tajkhorshid, Ferenc Molnar, James Phillips, and Klaus Schulten. Molecular dynamics study of bacteriorhodopsin and the purple membrane. *Journal of Physical Chemistry B*, 105:905–918, 2001.

[18] A.D. Becke. Density-Functional exchange-energy approximation with correct assymptotic behavior. *Phys. Rev. A*, 38:3098, (1988).

[19] R. J. Beckman, K. A. Baggerly, and M. D. McKay. Creating synthetic baseline populations. *Transportation Research Part A: Policy and Practice*, 30(6):415–429, 1996.

[20] Milind Bhandarkar, L. V. Kale, Eric de Sturler, and Jay Hoeflinger. Object-Based Adaptive Load Balancing for MPI Programs. In *Proceed-*

ings of the International Conference on Computational Science, San Francisco, CA, LNCS 2074, pages 108–117, May 2001.

[21] Abhinav Bhatele, Eric Bohm, and Laxmikant V. Kale. Optimizing communication for Charm++ applications by reducing network contention. *Concurrency and Computation: Practice and Experience,* 23(2):211–222, 2011.

[22] Abhinav Bhatele, Laxmikant V. Kale, and Sameer Kumar. Dynamic topology aware load balancing algorithms for molecular dynamics applications. In *23rd ACM International Conference on Supercomputing,* 2009.

[23] Scott Biersdorff, Chee Wai Lee, Allen D. Malony, and Laxmikant V. Kale. Integrated Performance Views in Charm ++: Projections Meets TAU. In *Proceedings of the 38th International Conference on Parallel Processing (ICPP),* pages 140–147, Vienna, Austria, September 2009.

[24] Keith Bisset, Ashwin Aji, Madhav Marathe, and Wu-chun Feng. High-performance biocomputing for simulating the spread of contagion over large contact networks. *BMC Genomics,* 13(Suppl 2):S3, 2012.

[25] Eric Bohm, Abhinav Bhatele, Laxmikant V. Kale, Mark E. Tuckerman, Sameer Kumar, John A. Gunnels, and Glenn J. Martyna. Fine Grained Parallelization of the Car-Parrinello ab initio MD Method on Blue Gene/L. *IBM Journal of Research and Development: Applications of Massively Parallel Systems,* 52(1/2):159–174, 2008.

[26] Kevin J. Bowers, Edmond Chow, Huafeng Xu, Ron O. Dror, Michael P. Eastwood, Brent A. Gregersen, John L. Klepeis, Istvan Kolossvary, Mark A. Moraes, Federico D. Sacerdoti, John K. Salmon, Yibing Shan, and David E. Shaw. Scalable algorithms for molecular dynamics simulations on commodity clusters. In *SC '06: Proceedings of the 2006 ACM/IEEE Conference on Supercomputing,* New York, NY, USA, 2006. ACM Press.

[27] BRAMS. http://www.cptec.inpe.br/brams, 2009.

[28] S. Browne, J. Dongarra, N. Garner, K. London, and P. Mucci. A scalable cross-platform infrastructure for application performance tuning using hardware counters. In *Proceedings of Supercomputing'00,* Dallas, Texas, 2000.

[29] Robert K. Brunner and Laxmikant V. Kalé. Adapting to load on workstation clusters. In *The Seventh Symposium on the Frontiers of Massively Parallel Computation,* pages 106–112. IEEE Computer Society Press, February 1999.

[30] Robert K. Brunner and Laxmikant V. Kalé. Handling application-induced load imbalance using parallel objects. In *Parallel and Distributed Computing for Symbolic and Irregular Applications*, pages 167–181. World Scientific Publishing, 2000.

[31] G. T. Camacho and M. Ortiz. Computational modeling of impact damage in brittle materials. *Int. J. Solids Struct.*, 33:2899–2938, 1996.

[32] R. Car and M. Parrinello. Unified approach for molecular dynamics and density functional theory. *Phys. Rev. Lett.*, 55:2471, (1985).

[33] C. Cavazzoni, G.L. Chiarotti, S. Scandolo, E. Tosatti, M. Bernasconi, and M. Parrinello. Superionic and Metallic States of Water and Ammonia at Giant Planet Conditions. *Science*, 283:44, (1999).

[34] Sayantan Chakravorty and L. V. Kale. A fault tolerant protocol for massively parallel machines. In *FTPDS Workshop for IPDPS 2004*. IEEE Press, 2004.

[35] Sayantan Chakravorty and L. V. Kale. A fault tolerance protocol with fast fault recovery. In *Proceedings of the 21st IEEE International Parallel and Distributed Processing Symposium*. IEEE Press, 2007.

[36] K. Channakeshava, K. Bisset, M. Marathe, A. Vullikanti, and S. Yardi. High performance scalable and expressive modeling environment to study mobile malware in large dynamic networks. In *Proceedings of 25th IEEE International Parallel & Distributed Processing Symposium*, 2011.

[37] Karthik Channakeshava, Deepti Chafekar, Keith Bisset, Anil Vullikanti, and Madhav Marathe. EpiNet: A simulation framework to study the spread of malware in wireless networks. In *SIMUTools09*. ICST Press, March 2009. Rome, Italy.

[38] K.L. Chung, Y.L. Huang, and Y.W. Liu. Efficient algorithms for coding Hilbert curve of arbitrary-sized image and application to window query. *Information Sciences*, 177(10):2130–2151, 2007.

[39] A.J. Cohen, Paula Mori-Sanchez, and Weitao Yang. Insights into current limitations of density functional theory. *Science*, 321:792, (2008).

[40] P. Colella, D.T. Graves, T.J. Ligocki, D.F. Martin, D. Modiano, D.B. Serafini, and B. Van Straalen. Chombo Software Package for AMR Applications Design Document, 2003. http://seesar.lbl.gov/anag/chombo/ChomboDesign-1.4.pdf.

[41] M. C.Payne, M.P. Teter, D.C. Allan, T.A. Arias, and J.D. Joannopoulos. Iterative minimization techniques for ab initio total-energy calculations: molecular dynamics and conjugate gradients. *Rev. Mod. Phys.*, 64:1045, (1992).

[42] T.A. Darden, D.M. York, and L.G. Pedersen. Particle mesh Ewald. An N·log(N) method for Ewald sums in large systems. *JCP*, 98:10089–10092, 1993.

[43] M. Davis, G. Efstathiou, C. S. Frenk, and S. D. M. White. The evolution of large-scale structure in a universe dominated by cold dark matter. *Astrophys. J.*, 292:371–394, May 1985.

[44] W. Dehnen. Towards optimal softening in three-dimensional N-body codes - I. Minimizing the force error. *MNRAS*, 324:273–291, June 2001.

[45] S.W. deLeeuw, J.W. Perram, and E.R. Smith. Simulation of Electrostatic Systems in Periodic Boundary Conditions. I. Lattice Sums and Dielectric Constants. *Proc. R. Soc. London A*, 373:27, 1980.

[46] Department of Computer Science, University of Illinois at Urbana-Champaign, Urbana, IL. *The CHARM (5.9) programming language manual*, 2006.

[47] Department of Computer Science, University of Illinois at Urbana-Champaign, Urbana, IL. *The CONVERSE programming language manual*, 2006.

[48] Jayant DeSouza and Laxmikant V. Kalé. MSA: Multiphase specifically shared arrays. In *Proceedings of the 17th International Workshop on Languages and Compilers for Parallel Computing*, West Lafayette, Indiana, USA, September 2004.

[49] K. Devine, B. Hendrickson, E. Boman, M. St. John, and C. Vaughan. Design of Dynamic Load-Balancing Tools for Parallel Applications. In *Proc. Intl. Conf. Supercomputing*, May 2000.

[50] Karen D. Devine, Erik G. Boman, Robert T. Heaphy, Bruce A. Hendrickson, James D. Teresco, Jamal Faik, Joseph E. Flaherty, and Luis G. Gervasio. New challenges in dynamic load balancing. *Appl. Numer. Math.*, 52(2–3):133–152, 2005.

[51] J. Diemand, M. Kuhlen, P. Madau, M. Zemp, B. Moore, D. Potter, and J. Stadel. Clumps and streams in the local dark matter distribution. *Nature*, 454:735–738, August 2008.

[52] H.-Q. Ding, N. Karasawa, and W. A. Goddard, III. The reduced cell multipole method for Coulomb interactions in periodic systems with million-atom unit cells. *Chemical Physics Letters*, 196:6–10, August 1992.

[53] P. Domingos and M. Richardson. Mining the Network Value of Customers. In *Proc. ACM KDD*, pages 57–61, 2001.

[54] Isaac Dooley. *Intelligent Runtime Tuning of Parallel Applications With Control Points*. PhD thesis, Dept. of Computer Science, University of Illinois, 2010. http://charm.cs.uiuc.edu/papers/DooleyPhD Thesis10.shtml.

[55] D. J. Earl and M.W. Deem. Parallel tempering: Theory, applications, and new perspectives. *Phys. Chem. Chem. Phys.*, 7:3910–3916, (2005).

[56] D. Easley and J. Kleinberg. *Networks, Crowds and Markets: Reasoning About A Highly Connected World*. Cambridge University Press, New York, NY, 2010.

[57] G. Efstathiou, M. Davis, S. D. M. White, and C. S. Frenk. Numerical techniques for large cosmological N-body simulations. *Astrophys. J. Supp.*, 57:241–260, February 1985.

[58] S.N. Eliane, E. Araújo, W. Cirne, G. Wagner, N. Oliveira, E.P. Souza, C.O. Galvão, and E.S. Martins. The SegHidro Experience: Using the Grid to Empower a Hydro-Meteorological Scientific Network. In *Proceedings of the First International Converence on e-Science and Grid Computing (e-Science/05)*, pages 64–71, 2005.

[59] S. Eubank, H. Guclu, V. S. Anil Kumar, M. Marathe, A. Srinivasan, Z. Toroczkai, and N. Wang. Modelling disease outbreaks in realistic urban social networks. *Nature*, 429:180–184, 2004.

[60] A. E. Evrard. Beyond N-body - 3D cosmological gas dynamics. *MNRAS*, 235:911–934, December 1988.

[61] P. P. Ewald. Die Berechnung optischer und elektrostatischer Gitterpotentiale. *Annalen der Physik*, 369:253–287, 1921.

[62] A. L. Fazenda, J. Panetta, P. Navaux, L. F. Rodrigues, D. M. Katsurayama, and L. F Motta. Escalabilidade de aplicação operacional em ambiente massivamente paralelo. In *Anais do X Simpósio em Sistemas Computacionais (WSCAD-SCC)*, pages 27–34, 2009.

[63] R.P. Feynman. *Statistical Mechanics*. Benjamin, Reading, (1972).

[64] B. Fitch, R. Germain, M. Mendell, J. Pitera, M. Pitman, A. Rayshubskiy, Y. Sham, F. Suits, W. Swope, T. Ward, Y. Zhestkov, and R. Zhou. Blue Matter, an application framework for molecular simulation on Blue Gene. *Journal of Parallel and Distributed Computing*, 63:759–773, 2003.

[65] Blake G. Fitch, Aleksandr Rayshubskiy, Maria Eleftheriou, T. J. Christopher Ward, Mark Giampapa, Michael C. Pitman, and Robert S. Germain. Molecular dynamics—blue matter: approaching the limits of concurrency for classical molecular dynamics. In *SC '06: Proceedings of the 2006 ACM/IEEE Conference on Supercomputing*, page 87, New York, NY, USA, 2006. ACM Press.

[66] I.T. Foster and B.R. Toonen. Load-balancing algorithms for climate models. In *Proceedings of Scalable High-Performance Computing Conference*, pages 674–681, 1994.

[67] Peter L. Freddolino, Anton S. Arkhipov, Steven B. Larson, Alexander McPherson, and Klaus Schulten. Molecular dynamics simulations of the complete satellite tobacco mosaic virus. *Structure*, 14:437–449, 2006.

[68] S.R. Freitas, K.M. Longo, M.A.F Silva Dias, R. Chatfield, P. Silva Dias, P. Artaxo, M.O. Andreae, G. Grell, L.F. Rodrigues, A. Fazenda, et al. The Coupled Aerosol and Tracer Transport model to the Brazilian developments on the Regional Atmospheric Modeling System (CATT-BRAMS). *Atmospheric Chemistry and Physics*, 9(8):2843–2861, 2009.

[69] C. S. Frenk, S. D. M. White, P. Bode, J. R. Bond, G. L. Bryan, R. Cen, H. M. P. Couchman, A. E. Evrard, N. Gnedin, A. Jenkins, A. M. Khokhlov, A. Klypin, J. F. Navarro, M. L. Norman, J. P. Ostriker, J. M. Owen, F. R. Pearce, U.-L. Pen, M. Steinmetz, P. A. Thomas, J. V. Villumsen, J. W. Wadsley, M. S. Warren, G. Xu, and G. Yepes. The Santa Barbara Cluster Comparison Project: A Comparison of Cosmological Hydrodynamics Solutions. *Astrophys. J.*, 525:554–582, November 1999.

[70] D. Frenkel and B. Smit. *Understanding Molecular Simulation*. Academic Press, 1996.

[71] George Karypis and Vipin Kumar. A fast and high quality multilevel scheme for partitioning irregular graphs. *SIAM J. Sci. Comput.*, 20(1):359–392, 1998.

[72] George Karypis and Vipin Kumar. Multilevel k-Way Partitioning Scheme for Irregular Graphs. *Journal of Parallel and Distributed Computing*, 48:96–129, 1998.

[73] T. C. Germann, K. Kadau, I. M. Longini, Jr., and C. A. Macken. Mitigation strategies for pandemic influenza in the United States. *Proc. of National Academy of Sciences*, 103(15):5935–5940, April 11, 2006.

[74] P. H. Geubelle and J. Baylor. Impact-Induced Delamination of Composites: A 2d Simulation. *Composites B*, 29(B):589–602, 1998.

[75] R. Gevaerd, S. R. Freitas, and K. M. Longo. Numerical simulation of biomass burning emission and trasportation during 1998 Roraima fires. In *Proceedings of International Conference on Southern Hemisphere Meteorology and Oceanography (ICSHMO) 8*, 2006.

[76] S. Ghan, X. Bian, A. Hunt, and A. Coleman. The thermodynamic influence of subgrid orography in a global climate model. *Climate Dynamics*, 20(1):31–44, 2002.

[77] S. Ghan and T. Shippert. Load balancing and scalability of a subgrid
      orography scheme in a global climate model. *International Journal of
      High Performance Computing Applications*, 19(3):237, 2005.

[78] D.S. Ginley and D. Cahen. *Fundamentals of Materials for Energy and
      Environmental Sustainability*. Cambridge University Press, Cambridge,
      UK.

[79] Filippo Gioachin and Laxmikant V. Kalé. Dynamic High-Level Scripting
      in Parallel Applications. In *Proceedings of the 23rd IEEE International
      Parallel and Distributed Processing Symposium (IPDPS)*, Rome, Italy,
      May 2009.

[80] Filippo Gioachin, Amit Sharma, Sayantan Chakravorty, Celso Mendes,
      Laxmikant V. Kale, and Thomas R. Quinn. Scalable cosmology simula-
      tions on parallel machines. In *VECPAR 2006, LNCS 4395, pp. 476-489*,
      2007.

[81] Filippo Gioachin, Gengbin Zheng, and Laxmikant V. Kalé. Debugging
      Large Scale Applications in a Virtualized Environment. In *Proceedings
      of the 23rd International Workshop on Languages and Compilers for
      Parallel Computing (LCPC2010)*, number 10-11, Houston, TX (USA),
      October 2010.

[82] Filippo Gioachin, Gengbin Zheng, and Laxmikant V. Kalé. Robust
      Record-Replay with Processor Extraction. In *PADTAD '10: Proceed-
      ings of the 8th Workshop on Parallel and Distributed Systems: Testing,
      Analysis, and Debugging*, pages 9–19. ACM, July 2010.

[83] E. Goldstein, A. Apolloni, B. Lewis, J. C. Miller, M. Macauley, S. Eu-
      bank, M. Lipsitch, and J. Wallinga. Distribution of vaccine/antivirals
      and the 'least spread line'; in a stratified population. *J. R. Soc. Inter-
      face*, 7(46):755–64, 2010.

[84] R. Gould. Collective action and network structure. *American Sociolog-
      ical Review*, 58:182–196, 1993.

[85] F. Governato, C. Brook, L. Mayer, A. Brooks, G. Rhee, J. Wadsley,
      P. Jonsson, B. Willman, G. Stinson, T. Quinn, and P. Madau. Bulgeless
      dwarf galaxies and dark matter cores from supernova-driven outflows.
      Nature, 463:203–206, January 2010.

[86] F. Governato, B. Willman, L. Mayer, A. Brooks, G. Stinson, O. Valen-
      zuela, J. Wadsley, and T. Quinn. Forming disc galaxies in ΛCDM sim-
      ulations. *MNRAS*, 374:1479–1494, February 2007.

[87] S. L . Graham, P. B. Kessler, and M. K. McKusick. GPROF: A call
      graph execution profiler. *SIGPLAN 1982 Symposium on Compiler Con-
      struction*, pages 120–126, June 1982.

[88] M. Granovetter. Threshold Models of Collective Behavior. *American J. Sociology*, 83(6):1420–1443, 1978.

[89] L. Greengard. *The rapid evaluation of potential fields in particle systems.* PhD thesis, MIT, Cambridge, MA, USA, 1988.

[90] G. Grell and D. Devenyi. A generalized approach to parameterizing convection combining ensemble and data assimilation techniques. *Geophysical Research Letters*, 29(14):38–1, 2002.

[91] G. Grimmett. *Percolation.* Springer, 1989.

[92] A. Gursoy, L.V. Kale, and S.P. Vanka. Unsteady fluid flow calculations using a machine independent parallel programming environment. In R. B. Pelz, A. Ecer, and J. Hauser, editors, *Parallel Computational Fluid Dynamics '92*, pages 175–185. North-Holland, 1993.

[93] A. Haldane and R. May. Systemic risk in banking ecosystems. *Nature*, 469:351–355, 2011.

[94] M. Elizabeth Halloran, Neil M. Ferguson, Stephen Eubank, Ira M. Longini, Derek A. T. Cummings, Bryan Lewis, Shufu Xu, Christophe Fraser, Anil Vullikanti, Timothy C. Germann, Diane Wagener, Richard Beckman, Kai Kadau, Chris Barrett, Catherine A. Macken, Donald S. Burke, and Philip Cooley. Modeling targeted layered containment of an influenza pandemic in the United States. *Proceedings of the National Academy of Sciences*, 105(12):4639–4644, March 2008.

[95] R. Halstead. Multilisp: A Language for Concurrent Symbolic Computation. *ACM Transactions on Programming Languages and Systems*, October 1985.

[96] Tsuyoshi Hamada, Tetsu Narumi, Rio Yokota, Kenji Yasuoka, Keigo Nitadori, and Makoto Taiji. 42 tflops hierarchical n-body simulations on gpus with applications in both astrophysics and turbulence. In *Proceedings of the Conference on High Performance Computing Networking, Storage and Analysis*, SC '09, pages 62:1–62:12, New York, NY, USA, 2009. ACM.

[97] Richard Hamming. *Numerical Analysis for Scientists and Engineers.* Dover Publications, March 1973.

[98] K. Heitmann, P. M. Ricker, M. S. Warren, and S. Habib. Robustness of Cosmological Simulations. I. Large-Scale Structure. *Astrophys. J. Supp.*, 160:28–58, September 2005.

[99] L. Hernquist, F. R. Bouchet, and Y. Suto. Application of the Ewald method to cosmological N-body simulations. *Astrophys. J. Supp.*, 75:231–240, February 1991.

[100] L. Hernquist and N. Katz. TREESPH - A unification of SPH with the hierarchical tree method. *Astrophys. J. Supp.*, 70:419–446, June 1989.

[101] D. Hilbert. Über die stetige abbildung einer linie auf ein flächenstück. *Mathematische Annalen*, 38:459–460, 1891.

[102] R. W. Hockney and J. W. Eastwood. *Computer Simulation Using Particles.* New York: McGraw-Hill, 1981.

[103] P. Hohenberg and W. Kohn. Inhomogeneous electron gas. *Phys. Rev.*, 136:B864, 1964.

[104] Chao Huang and Laxmikant V. Kale. Charisma: Orchestrating migratable parallel objects. In *Proceedings of IEEE International Symposium on High Performance Distributed Computing (HPDC)*, July 2007.

[105] Chao Huang, Orion Lawlor, and L. V. Kalé. Adaptive MPI. In *Proceedings of the 16th International Workshop on Languages and Compilers for Parallel Computing (LCPC 2003), LNCS 2958*, pages 306–322, College Station, Texas, October 2003.

[106] Chao Huang, Gengbin Zheng, Sameer Kumar, and Laxmikant V. Kalé. Performance Evaluation of Adaptive MPI. In *Proceedings of ACM SIG-PLAN Symposium on Principles and Practice of Parallel Programming 2006*, March 2006.

[107] J. JáJá. *An Introduction to Parallel Algorithms.* Addison Wesley Longman Publishing Co., Inc., Redwood City, CA, USA, 1992.

[108] Pritish Jetley, Filippo Gioachin, Celso Mendes, Laxmikant V. Kale, and Thomas R. Quinn. Massively parallel cosmological simulations with ChaNGa. In *Proceedings of IEEE International Parallel and Distributed Processing Symposium 2008*, 2008.

[109] Pritish Jetley, Lukasz Wesolowski, Filippo Gioachin, Laxmikant V. Kalé, and Thomas R. Quinn. Scaling hierarchical n-body simulations on gpu clusters. In *Proceedings of the 2010 ACM/IEEE International Conference for High Performance Computing, Networking, Storage and Analysis*, SC '10, Washington, DC, USA, 2010. IEEE Computer Society.

[110] Xiangmin Jiao, Gengbin Zheng, Phillip A. Alexander, Michael T. Campbell, Orion S. Lawlor, John Norris, Andreas Haselbacher, and Michael T. Heath. A system integration framework for coupled multiphysics simulations. *Engineering with Computers*, 22(3):293–309, 2006.

[111] J. M. Jiménez, B. L. Lewis, and S. Eubank. Hospitals as complex social systems: Agent-based simulation of hospital-acquired infections. In *Proceedings of 2nd International Conference on Complex Sciences: Theory and Applications*, 2012.

[112] John A. Board, Jr., Laxmikant V. Kale, Klaus Schulten, Robert D. Skeel, and Tamar Schlick. Modeling biomolecules: Large sclaes, longer durations. *IEEE Computational Science & Engineering*, 1:19–30, Winter 1994.

[113] Rashmi Jyothi, Orion Sky Lawlor, and L. V. Kale. Debugging support for Charm++. In *PADTAD Workshop for IPDPS 2004*, page 294. IEEE Press, 2004.

[114] L. V. Kale. Application oriented and computer science centered research. HPC Workshops, pages 98–105, 1994.

[115] L. V. Kale and Milind Bhandarkar. Structured dagger: A Coordination Language for Message-Driven Programming. In *Proceedings of Second International Euro-Par Conference*, volume 1123-1124 of *Lecture Notes in Computer Science*, pages 646–653, September 1996.

[116] L. V. Kale and Sanjeev Krishnan. A comparison based parallel sorting algorithm. In *Proceedings of the 22nd International Conference on Parallel Processing*, pages 196–200, St. Charles, IL, August 1993.

[117] L. V. Kale and Sanjeev Krishnan. Charm++: Parallel Programming with Message-Driven Objects. In Gregory V. Wilson and Paul Lu, editors, *Parallel Programming using C++*, pages 175–213. MIT Press, 1996.

[118] L. V. Kale, B. H. Richards, and T. D. Allen. Efficient parallel graph coloring with prioritization. In *Lecture Notes in Computer Science*, volume 1068, pages 190–208. Springer-Verlag, August 1995.

[119] Laxmikant Kale, Anshu Arya, Abhinav Bhatele, Abhishek Gupta, Nikhil Jain, Pritish Jetley, Jonathan Lifflander, Phil Miller, Yanhua Sun, Ramprasad Venkataraman, Lukasz Wesolowski, and Gengbin Zheng. Charm++ for productivity and performance: A submission to the 2011 HPC class II challenge. Technical Report 11-49, Parallel Programming Laboratory, November 2011.

[120] Laxmikant Kale, Anshu Arya, Nikhil Jain, Akhil Langer, Jonathan Lifflander, Harshitha Menon, Xiang Ni, Yanhua Sun, Ehsan Totoni, Ramprasad Venkataraman, and Lukasz Wesolowski. Migratable objects + active messages + adaptive runtime = productivity + performance a submission to 2012 HPC class II challenge. Technical Report 12-47, Parallel Programming Laboratory, November 2012.

[121] Laxmikant Kale, Robert Skeel, Milind Bhandarkar, Robert Brunner, Attila Gursoy, Neal Krawetz, James Phillips, Aritomo Shinozaki, Krishnan Varadarajan, and Klaus Schulten. NAMD2: Greater scalability for parallel molecular dynamics. *Journal of Computational Physics*, 151:283–312, 1999.

[122] Laxmikant V. Kale. The virtualization model of parallel programming: Runtime optimizations and the state of art. In *LACSI 2002*, Albuquerque, October 2002.

[123] Laxmikant V. Kale. Performance and productivity in parallel programming via processor virtualization. In *Proc. of the First Intl. Workshop on Productivity and Performance in High-End Computing (at HPCA 10)*, Madrid, Spain, February 2004.

[124] Laxmikant V. Kale, Sameer Kumar, Gengbin Zheng, and Chee Wai Lee. Scaling molecular dynamics to 3000 processors with projections: A performance analysis case study. In *Terascale Performance Analysis Workshop, International Conference on Computational Science(ICCS)*, Melbourne, Australia, June 2003.

[125] Laxmikant V. Kale, Gengbin Zheng, Chee Wai Lee, and Sameer Kumar. Scaling applications to massively parallel machines using projections performance analysis tool. In *Future Generation Computer Systems Special Issue on: Large-Scale System Performance Modeling and Analysis*, volume 22, pages 347–358, February 2006.

[126] L.V. Kale and S. Krishnan. CHARM++: A Portable Concurrent Object Oriented System Based on C++. In A. Paepcke, editor, *Proceedings of OOPSLA'93*, pages 91–108. ACM Press, September 1993.

[127] L.V. Kale and Amitabh Sinha. Projections: A preliminary performance tool for charm. In *Parallel Systems Fair, International Parallel Processing Symposium*, pages 108–114, Newport Beach, CA, April 1993.

[128] S.A. Kalogirou. *Solar Energy Engineering: Processes and Systems*. Academic Press, Waltham, MA, USA.

[129] George Karypis and Vipin Kumar. METIS: Unstructured graph partitioning and sparse matrix ordering system. University of Minnesota, 1995.

[130] George Karypis and Vipin Kumar. Parallel multilevel k-way partitioning scheme for irregular graphs. In *Supercomputing '96: Proceedings of the 1996 ACM/IEEE conference on Supercomputing (CDROM)*, page 35, 1996.

[131] Amal Kasry, Marcelo A. Kuroda, Glenn J. Martyna, George S. Tulevski, and Ageeth A. Bol. Chemical doping of large-area stacked graphene films for use as transparent, conducting electrodes. *ACS Nano*, 4(7):3839–3844, 2010.

[132] D. Kempe, J. Kleinberg, and E. Tardos. Maximizing the Spread of Influence through a Social Network. In *Proc. ACM KDD*, pages 137–146, 2003.

[133] D. Kempe, J. Kleinberg, and E. Tardos. Influential Nodes in a Diffusion Model for Social Networks. In *Proc. ICALP*, pages 1127–1138, 2005.

[134] C.H. Koelbel, D.B. Loveman, R.S. Schreiber, G.L. Steele, Jr., and M.E. Zosel. *The High Performance Fortran Handbook*. MIT Press, 1994.

[135] W. Kohn and L.J. Sham. Self-consistent equations including exchange and correlation effects. *Phys. Rev.*, 140:A1133, 1965.

[136] C. Koziar, R. Reilein, and G. Runger. Load imbalance aspects in atmosphere simulations. *International Journal of Computational Science and Engineering*, 1(2):215–225, 2005.

[137] Sanjeev Krishnan and L. V. Kale. A parallel adaptive fast multipole algorithm for n-body problems. In *Proceedings of the International Conference on Parallel Processing*, pages III 46–III 50, August 1995.

[138] Rick Kufrin. Perfsuite: An Accessible, Open Source Performance Analysis Environment for Linux. In *Proceedings of the Linux Cluster Conference*, 2005.

[139] Sameer Kumar. *Optimizing Communication for Massively Parallel Processing*. PhD thesis, University of Illinois at Urbana-Champaign, May 2005.

[140] Sameer Kumar, Chao Huang, Gheorghe Almasi, and Laxmikant V. Kalé. Achieving strong scaling with NAMD on Blue Gene/L. In *Proceedings of IEEE International Parallel and Distributed Processing Symposium 2006*, April 2006.

[141] V. Kumar. *Introduction to Parallel Computing*. Addison Wesley Longman Publishing Co., Inc., Boston, MA, USA, 2002.

[142] Akhil Langer, Jonathan Lifflander, Phil Miller, Kuo-Chuan Pan, Laxmikant V. Kale, and Paul Ricker. Scalable Algorithms for Distributed-Memory Adaptive Mesh Refinement. In *Proceedings of the 24th International Symposium on Computer Architecture and High Performance Computing (SBAC-PAD 2012)*. New York, USA, October 2012.

[143] Ilya Lashuk, Aparna Chandramowlishwaran, Harper Langston, Tuan-Anh Nguyen, Rahul Sampath, Aashay Shringarpure, Richard Vuduc, Lexing Ying, Denis Zorin, and George Biros. A massively parallel adaptive fast-multipole method on heterogeneous architectures. In *SC '09: Proceedings of the Conference on High Performance Computing Networking, Storage and Analysis*, pages 1–12, New York, NY, USA, 2009. ACM.

[144] Orion Lawlor, Sayantan Chakravorty, Terry Wilmarth, Nilesh Choudhury, Isaac Dooley, Gengbin Zheng, and Laxmikant Kale. Parfum: A parallel framework for unstructured meshes for scalable dynamic physics applications. *Engineering with Computers*, 22(3-4):215–235, September 2006.

[145] Orion Lawlor, Hari Govind, Isaac Dooley, Michael Breitenfeld, and Laxmikant Kale. Performance degradation in the presence of subnormal floating-point values. In *Proceedings of the International Workshop on Operating System Interference in High Performance Applications*, September 2005.

[146] Orion Sky Lawlor. *Impostors for Parallel Interactive Computer Graphics*. PhD thesis, University of Illinois at Urbana-Champaign, December 2004.

[147] Orion Sky Lawlor and L. V. Kalé. Supporting dynamic parallel object arrays. *Concurrency and Computation: Practice and Experience*, 15:371–393, 2003.

[148] D. Lea and W. Gloger. A memory allocator. `http://web.mit.edu/ sage/export/singular-3-0-4-3.old/omalloc/Misc/dlmalloc/ malloc.ps`, 2000.

[149] C. Lee, W. Yang, and R.G. Parr. Development of the Calle-Salvetti correlation energy into a functional of the electron density. *Phys. Rev. B*, 37:785, (1988).

[150] Chee Wai Lee. *Techniques in Scalable and Effective Parallel Performance Analysis*. PhD thesis, Department of Computer Science, University of Illinois, Urbana-Champaign, December 2009.

[151] J. K. Lenstra, D. B. Shmoys, and E. Tardos. Approximation algorithms for scheduling unrelated parallel machines. *Math. Program.*, 46(3):259–271, 1990.

[152] J. Leskovec, L. Adamic, and B. Huberman. The Dynamics of Viral Marketing. *ACM Trans. on the Web*, 1(1), 2007.

[153] J.R. Levine. *Linkers and Loaders*. Morgan-Kauffman, 2000.

[154] Bryan Lewis, Stephen Eubank, Allyson M Abrams, and Ken Kleinman. In silico surveillance: Evaluating outbreak detection with simulation models. *BMC Medical Informatics and Decision Making*, 13(1):12, January 2013.

[155] G. F. Lewis, A. Babul, N. Katz, T. Quinn, L. Hernquist, and D. H. Weinberg. The Effects of Gasdynamics, Cooling, Star Formation, and Numerical Resolution in Simulations of Cluster Formation. *Astrophys. J.*, 536:623–644, June 2000.

[156] X. Li, W. Cai, J. An, S. Kim, J. Nah, D. Yang, R. Piner, A. Velamakanni, I. Jung, E. Tutuc, S.K. Banerjee, L. Colombo, and R.S. Ruoff. Large-Area Synthesis of High-Quality and Uniform Graphene Films on Copper Foils. *Science*, 324:1312, (2009).

[157] X. Liu and G. Schrack. Encoding and decoding the Hilbert order. *Software, Practice & Experience*, 26(12):1335–1346, 1996.

[158] Kwan-Liu Ma, Greg Schussman, Brett Wilson, Kwok Ko, Ji Qiang, and Robert Ryne. Advanced visualization technology for terascale particle accelerator simulations. In *Supercomputing '02: Proceedings of the 2002 ACM/IEEE Conference on Supercomputing*, pages 1–11, Los Alamitos, CA, USA, 2002. IEEE Computer Society Press.

[159] Paulo W. C. Maciel and Peter Shirley. Visual navigation of large environments using textured clusters. In *Proceedings of the 1995 Symposium on Interactive 3D Graphics*, pages 95–ff. ACM Press, 1995.

[160] Sandhya Mangala, Terry Wilmarth, Sayantan Chakravorty, Nilesh Choudhury, Laxmikant V. Kale, and Philippe H. Geubelle. Parallel adaptive simulations of dynamic fracture events. *Engineering with Computers*, 24:341–358, December 2007.

[161] Achla Marathe, Bryan Lewis, Christopher Barrett, Jiangzhuo Chen, Madhav Marathe, Stephen Eubank, and Yifei Ma. Comparing effectiveness of top-down and bottom-up strategies in containing influenza. *PloS one*, 6(9):e25149, 2011.

[162] Achla Marathe, Bryan Lewis, Jiangzhuo Chen, and Stephen Eubank. Sensitivity of household transmission to household contact structure and size. *PloS one*, 6(8):e22461, 2011.

[163] Dominik Marx, Mark E. Tuckerman, and M. Parrinello. The nature of the hydrated excess proton in water. *Nature*, 601:397, (1999).

[164] L. Mayer, T. Quinn, J. Wadsley, and J. Stadel. Formation of Giant Planets by Fragmentation of Protoplanetary Disks. *Science*, 298:1756–1759, November 2002.

[165] M. McPherson, L. Smith-Lovin, and J. Cook. Birds of a Feather: Homophily in Social Networks. *Annual Review of Sociology*, 27:415–444, 2001.

[166] Vikas Mehta. LeanMD: A Charm++ framework for high performance molecular dynamics simulation on large parallel machines. Master's thesis, University of Illinois at Urbana-Champaign, 2004.

[167] Chao Mei, Yanhua Sun, Gengbin Zheng, Eric J. Bohm, Laxmikant V. Kale, James C.Phillips, and Chris Harrison. Enabling and scaling

biomolecular simulations of 100 million atoms on petascale machines with a multicore-optimized message-driven runtime. In *Proceedings of the 2011 ACM/IEEE Conference on Supercomputing*, Seattle, WA, November 2011.

[168] Chao Mei, Gengbin Zheng, Filippo Gioachin, and Laxmikant V. Kale. Optimizing a Parallel Runtime System for Multicore Clusters: A case study. In *TeraGrid'10*, number 10-13, Pittsburgh, PA, USA, August 2010.

[169] Esteban Meneses, Greg Bronevetsky, and Laxmikant V. Kale. Dynamic load balance for optimized message logging in fault tolerant hpc applications. In *IEEE International Conference on Cluster Computing (Cluster) 2011*, September 2011.

[170] Esteban Meneses, Celso L. Mendes, and Laxmikant V. Kale. Team-based message logging: Preliminary results. In *3rd Workshop on Resiliency in High Performance Computing (Resilience) in Clusters, Clouds, and Grids (CCGRID 2010)*, May 2010.

[171] J. Michalakes. MM90: A scalable parallel implementation of the Penn State/NCAR Mesoscale Model (MM5). *Parallel Computing*, 23(14):2173–2186, 1997.

[172] John Michalakes, Josh Hacker, Richard Loft, Michael O. McCracken, Allan Snavely, Nicholas J. Wright, Tom Spelce, Brent Gorda, and Robert Walkup. Wrf nature run. In *Proceedings of SuperComputing*, pages 1–6, Los Alamitos, CA, USA, 2007. IEEE Computer Society.

[173] Phil Miller, Aaron Becker, and Laxmikant Kale. Using shared arrays in message-driven parallel programs. *Parallel Computing*, 38(12):66–74, 2012.

[174] J. Minkel. The 2003 northeast blackout–five years later. *Scientific American*, 2008. 13 August 2008, http://www.scientificamerican.com/article.cfm?id=2003-blackout-five-years-later.

[175] M. Levy. Universal variational functionals of electron densities, first-order density matrices, and natural spin-orbitals and solution of the v-representability problem. *Proc. Natl. Acad. Sci. U.S.A.*, 76:6062, (1979).

[176] P. R. Monge and N. S. Contractor. *Theories of Communication Networks*. Oxford University Press, USA, 2003.

[177] B. Moore, F. Governato, T. Quinn, J. Stadel, and G. Lake. Resolving the Structure of Cold Dark Matter Halos. *Astrophys. J. Lett.*, 499:L5–+, May 1998.

[178] E. Moretti. Social learning and peer effects in consumption: Evidence from movie sales. *Review of Economic Studies*, 78:356–393, 2011.

[179] H. Mortveit and C. Reidys. *An Introduction to Sequential Dynamical Systems.* Springer, New York, NY, 2007.

[180] Martin Mundhenk, Judy Goldsmith, Christopher Lusena, and Eric Allender. Complexity of finite-horizon Markov decision process problems. *JACM*, 47(4):681–720, July 2000.

[181] National Institutes of Health, 2009. `http://www.nigms.nih.gov/Initiatives/MIDAS`.

[182] J. F. Navarro, C. S. Frenk, and S. D. M. White. A Universal Density Profile from Hierarchical Clustering. *Astrophys. J.*, 490:493, December 1997.

[183] NDSSL. Synthetic data products for societal infrastructures and proto-populations: Data set 2.0. Technical Report NDSSL-TR-07-003, NDSSL, Virginia Polytechnic Institute and State University, Blacksburg, VA, 2007.

[184] M. Newman. The structure and function of complex networks. *SIAM Review*, 45, 2003.

[185] M.E. Newman. Spread of epidemic disease on networks. *Phys. Rev. E*, 2002.

[186] D.M. Newns, B.G. Elmegreen, X.-H. Liu, and G.J. Martyna. High Response Piezoelectric and Piezoresistive Materials for Fast, Low Voltage Switching: Simulation and Theory of Transduction Physics at the Nanometer-Scale. *Adv. Mat.*, 24:3672, 2012.

[187] D.M. Newns, B.G. Elmegreen, X.-H. Liu, and G.J. Martyna. The piezoelectronic transistor: A nanoactuator-based post-CMOS digital switch with high speed and low power. *MRS Bulletin*, 37:1071, 2012.

[188] D.M. Newns, J.A. Misewich, A. Gupta C.C. Tsuei, B.A. Scott, and A. Schrott. Mott Transition Field Effect Transistor. *Appl. Phys. Lett.*, 73:780, (1998).

[189] R. Nistor, D.M. Newns, and G.J. Martyna. Understanding the doping mechanism in graphene-based electronics: The role of chemistry. *ACS Nano*, 5:3096, (2011).

[190] A. Odell, A. Delin, B. Johansson, N. Bock, M. Challacombe, and A. M. N. Niklasson. Higher-order symplectic integration in Born–Oppenheimer molecular dynamics. *J. Chem. Phys.*, 131:244106, (2009).

[191] Committee on Modeling Community Containment for Pandemic Influenza and Institute of Medicine. *Modeling Community Containment for Pandemic Influenza: A Letter Report.* The National Academies Press, Washington D.C., 2006.

[192] Ehsan Totoni, Osman Sarood, Phil Miller, and L. V. Kale. 'Cool' Load
      Balancing for High Performance Computing Data Centers. In *IEEE
      Transactions on Computer - SI (Energy Efficient Computing)*, September 2012.

[193] J. P. Ostriker and P. J. E. Peebles. A Numerical Study of the Stability
      of Flattened Galaxies: Or, Can Cold Galaxies Survive? *Astrophys. J.*,
      186:467–480, December 1973.

[194] Douglas Z. Pan and Mark A. Linton. Supporting reverse execution for
      parallel programs. *SIGPLAN Not.*, 24(1):124–129, 1989.

[195] J. P. Perdew, K. Burke, and M. Ernzerhof. Generalized Gradient Approximation Made Simple. *Phys. Rev. B*, 77:386, (1996).

[196] P. Perzyna. Fundamental problems in viscoplasticity. *Advances in Applied Mechanics*, 9(C):243–377, 1966.

[197] James C. Phillips, Gengbin Zheng, Sameer Kumar, and Laxmikant V.
      Kalé. NAMD: Biomolecular simulation on thousands of processors. In
      Proceedings of the 2002 ACM/IEEE Conference on Supercomputing,
      pages 1–18, Baltimore, MD, September 2002.

[198] Planck Collaboration, N. Aghanim, M. Arnaud, M. Ashdown, et al.
      Planck intermediate results. I. Further validation of new Planck clusters
      with XMM-Newton. *Astronomy and Astrophysics*, 543:A102, July 2012.

[199] S. J. Plimpton and B. A. Hendrickson. A new parallel method for
      molecular-dynamics simulation of macromolecular systems. *J. Comp.
      Chem.*, 17:326–337, 1996.

[200] Steve Plimpton. Fast parallel algorithms for short-range molecular dynamics. *J. Comput. Phys.*, 117(1):1–19, 1995.

[201] C. Power, J. F. Navarro, A. Jenkins, C. S. Frenk, S. D. M. White,
      V. Springel, J. Stadel, and T. Quinn. The inner structure of ΛCDM
      haloes - I. A numerical convergence study. *Monthly Notices of the Royal
      Astronomical Society*, 338:14–34, January 2003.

[202] D. Reed, J. Gardner, T. Quinn, J. Stadel, M. Fardal, G. Lake, and
      F. Governato. Evolution of the mass function of dark matter haloes.
      MNRAS, 346:565–572, December 2003.

[203] D.K. Remler and P.A. Madden. Molecular dynamics without effective
      potentials via the Car-Parrinello approach. *Mol. Phys.*, 70:921, (1990).

[204] E. R. Rodrigues, P. O. A. Navaux, J. Panetta, and C. L. Mendes. A
      new technique for data privatization in user-level threads and its use in
      parallel applications. In *ACM 25th Symposium on Applied Computing
      (SAC), Sierre, Switzerland*, 2010.

[205] Eduardo R. Rodrigues, Philippe O. A. Navaux, Jairo Panetta, Alvaro Fazenda, Celso L. Mendes, and Laxmikant V. Kalé. A comparative analysis of load balancing algorithms applied to a weather forecast model. In *Proceedings of 22nd IEEE International Symposium on Computer Architecture and High Performance Computing*, Petrópolis - Brazil, 2010.

[206] Eduardo R. Rodrigues, Philippe O. A. Navaux, Jairo Panetta, Celso L. Mendes, and Laxmikant V. Kalé. Optimizing an MPI weather forecasting model via processor virtualization. In *Proceedings of IEEE International Conference on High Performance Computing (HiPC 2010)*, Goa - India, 2010.

[207] D. Romero, B. Meeder, and J. Kleinberg. Differences in the Mechanics of Information Diffusion across Topics: Idioms, Political Hashtags, and Complex Contagion on Twitter. In *Proceedings of the 20th International World Wide Web Conference (WWW 2011)*, 2011.

[208] Michiel Ronsse and Koen De Bosschere. RecPlay: A fully integrated practical record/replay system. *ACM Trans. Comput. Syst.*, 17(2):133–152, 1999.

[209] H.G. Rotithor. Taxonomy of dynamic task scheduling schemes in distributed computing systems. In *Proceedings of IEE: Computers and Digital Techniques*, volume 141, pages 1–10, 1994.

[210] J. J. Ruan, T. R. Quinn, and A. Babul. The Observable Thermal and Kinetic Sunyaev-Zel'dovich Effect in Merging Galaxy Clusters. *ArXiv e-prints*, April 2013.

[211] Ruth Rutter. Run-length encoding on graphics hardware. Master's thesis, University of Alaska at Fairbanks, 2011.

[212] J. K. Salmon and M. S. Warren. Skeletons from the treecode closet. *Journal of Computational Physics*, 111:136–155, March 1994.

[213] Yanhua Sun, Sameer Kumar and L. V. Kale. Acceleration of an Asynchronous Message Driven Programming Paradigm on IBM Blue Gene/Q. In *Proceedings of 26th IEEE International Parallel and Distributed Processing Symposium (IPDPS)*, Boston, USA, May 2013.

[214] Osman Sarood and Laxmikant V. Kalé. A 'cool' load balancer for parallel applications. In *Proceedings of the 2011 ACM/IEEE Conference on Supercomputing*, Seattle, WA, November 2011.

[215] Martin Schulz, Jim Galarowicz, Don Maghrak, William Hachfeld, David Montoya, and Scott Cranford. Open|speedshop: An open source infrastructure for parallel performance analysis. *Scientific Programming*, 16(2-3):105–121, 2008.

[216] Melih Sener, Johan Strumpfer, John A. Timney, Arvi Freiberg, C. Neil Hunter, and Klaus Schulten. Photosynthetic vesicle architecture and constraints on efficient energy harvesting. *Biophysical Journal*, 99:67–75, 2010.

[217] D. Shakhvorostov, R.A. Nistor, L. Krusin-Elbaum, G.J. Martyna, D.M. Newns, B.G. Elmegreen, X. Liu, Z.E. Hughesa, S. Paul, C. Cabral, S. Raoux, D.B. Shrekenhamerd, D.N. Basovd, Y. Songe, and M.H. Mueser. Evidence for electronic gap-driven metal-semiconductor transition in phase-change materials. *PNAS*, 106:10907–10911, (2009).

[218] S. Shende and A. D. Malony. The TAU Parallel Performance System. *International Journal of High Performance Computing Applications*, 20(2):287–331, Summer 2006.

[219] S.A. Shevlin, A. Curioni, and W. Andreoni. Ab initio Design of High-k Dielectrics: $La_xY_{1-x}AlO_3$. *Phys. Rev. Lett.*, 94:146401, (2005).

[220] S. Shingu, H. Takahara, H. Fuchigami, M. Yamada, Y. Tsuda, W. Ohfuchi, Y. Sasaki, K. Kobayashi, T. Hagiwara, S. Habata, et al. A 26.58 tflops global atmospheric simulation with the spectral transform method on the earth simulator. In *Proceedings of the 2002 ACM/IEEE Conference on Supercomputing*, pages 1–19. IEEE Computer Society Press, 2002.

[221] D. Siegel. Social networks and collective action. *American Journal of Political Science*, 53:122–138, 2009.

[222] A. Sinha and L.V. Kalé. Information Sharing Mechanisms in Parallel Programs. In H.J. Siegel, editor, *Proceedings of the 8th International Parallel Processing Symposium*, pages 461–468, Cancun, Mexico, April 1994.

[223] Marc Snir. A note on n-body computations with cutoffs. *Theory of Computing Systems*, 37:295–318, 2004.

[224] Edgar Solomonik and Laxmikant V. Kale. Highly Scalable Parallel Sorting. In *Proceedings of the 24th IEEE International Parallel and Distributed Processing Symposium (IPDPS)*, April 2010.

[225] R. Souto, R.B. Avila, P.O.A. Navaux, M.X. Py, N. Maillard, T. Diverio, H.C. Velho, S. Stephany, A.J. Preto, J. Panetta, et al. Processing mesoscale climatology in a grid environment. In *Proceedings of the Seventh IEEE International Symposium on Cluster Computing and the Grid–CCGrid*, 2007.

[226] V. Springel. The cosmological simulation code GADGET-2. *MNRAS*, 364:1105–1134, December 2005.

[227] V. Springel, J. Wang, M. Vogelsberger, A. Ludlow, A. Jenkins, A. Helmi, J. F. Navarro, C. S. Frenk, and S. D. M. White. The Aquarius Project: The subhaloes of galactic haloes. *MNRAS*, 391:1685–1711, December 2008.

[228] V. Springel, S. D. M. White, A. Jenkins, C. S. Frenk, N. Yoshida, L. Gao, J. Navarro, R. Thacker, D. Croton, J. Helly, J. A. Peacock, S. Cole, P. Thomas, H. Couchman, A. Evrard, J. Colberg, and F. Pearce. Simulations of the formation, evolution and clustering of galaxies and quasars. *Nature*, 435:629–636, June 2005.

[229] J. Stadel, D. Potter, B. Moore, J. Diemand, P. Madau, M. Zemp, M. Kuhlen, and V. Quilis. Quantifying the heart of darkness with GHALO - a multibillion particle simulation of a galactic halo. *MNRAS*, 398:L21–L25, September 2009.

[230] J. G. Stadel. *Cosmological N-body Simulations and their Analysis*. PhD thesis, Department of Astronomy, University of Washington, March 2001.

[231] Yanhua Sun, Gengbin Zheng, Chao Mei Eric J. Bohm, Terry Jones, Laxmikant V. Kalé, and James C. Phillips. Optimizing fine-grained communication in a biomolecular simulation application on cray xk6. In *Proceedings of the 2012 ACM/IEEE Conference on Supercomputing*, Salt Lake City, Utah, November 2012.

[232] Yanhua Sun, Gengbin Zheng, L. V. Kale, Terry R. Jones, and Ryan Olson. A uGNI-based Asynchronous Message-Driven Runtime System for Cray Supercomputers with Gemini Interconnect. In *Proceedings of 26th IEEE International Parallel and Distributed Processing Symposium (IPDPS)*, Shanghai, China, May 2012.

[233] Emad Tajkhorshid, Aleksij Aksimentiev, Ilya Balabin, Mu Gao, Barry Isralewitz, James C. Phillips, Fangqiang Zhu, and Klaus Schulten. Large scale simulation of protein mechanics and function. In Frederic M. Richards, David S. Eisenberg, and John Kuriyan, editors, *Advances in Protein Chemistry*, volume 66, pages 195–247. Elsevier Academic Press, New York, 2003.

[234] Emad Tajkhorshid, Peter Nollert, Morten Ø. Jensen, Larry J. W. Miercke, Joseph O'Connell, Robert M. Stroud, and Klaus Schulten. Control of the selectivity of the aquaporin water channel family by global orientational tuning. *Science*, 296:525–530, 2002.

[235] Claudia Taylor, Achla Marathe, and Richard Beckman. Same influenza vaccination strategies but different outcomes across US cities? *International Journal of Infectious Diseases*, 14(9):e792–e795, 2010.

[236] T.N. Theis and P.M. Solomon. In quest of the "Next Switch": Prospects for Greatly Reduced Power Dissipation in a Successor to the Silicon Field-Effect Transistor. *Proceedings of the IEEE*, 98:2005, 2010.

[237] G.J. Tripoli and W.R. Cotton. The Colorado State University three-dimensional cloud/mesoscale model. Technical Report 3, Atmos, 1982.

[238] M. Tuckerman, G. Martyna, M.L. Klein, and B.J. Berne. Efficient Molecular Dynamics and Hybrid Monte Carlo Algorithms for Path Integrals. *J. Chem. Phys.*, 99:2796, 1993.

[239] Ramkumar V. Vadali, Yan Shi, Sameer Kumar, L. V. Kale, Mark E. Tuckerman, and Glenn J. Martyna. Scalable fine-grained parallelization of plane-wave-based ab initio molecular dynamics for large super-computers. *Journal of Comptational Chemistry*, 25(16):2006–2022, Oct. 2004.

[240] J. W. Wadsley, J. Stadel, and T. Quinn. Gasoline: A flexible, parallel implementation of TreeSPH. *New Astronomy*, 9:137–158, February 2004.

[241] R.L. Walko, L.E. Band, J. Baron, T.G.F. Kittel, R. Lammers, T.J. Lee, D. Ojima, R.A. Pielke Sr., C. Taylor, C. Tague, et al. Coupled atmosphere–biophysics–hydrology models for environmental modeling. *Journal of Applied Meteorology*, 39(6), 2000.

[242] Yuhe Wang and John Killough. A new approach to load balance for parallel compositional simulation based on reservoir model over-decomposition. In *2013 SPE Reservoir Simulation Symposium*, 2013.

[243] M. S. Warren and J. K. Salmon. A parallel hashed oct-tree n-body algorithm. In *Proceedings of the 1993 ACM/IEEE Conference on Supercomputing*, Supercomputing '93, pages 12–21, New York, NY, USA, 1993. ACM.

[244] M.S. Warren, J.K. Salmon, D.J. Becker, M.P. Goda, T. Sterling, and W. Winckelmans. Pentium pro inside: I. A treecode at 430 gigaflops on asci red, ii. Price/performance of $50/mflop on loki and hyglac. In *Supercomputing, ACM/IEEE 1997 Conference*, page 61, Nov. 1997.

[245] S. D. M. White, C. S. Frenk, and M. Davis. Clustering in a neutrino-dominated universe. *Astrophys. J. Lett.*, 274:L1–L5, November 1983.

[246] X.-P. Xu and A. Needleman. Numerical simulation of fast crack growth in brittle solids. *Journal of the Mechanics and Physics of Solids*, 42:1397–1434, 1994.

[247] M. Xue, K.K. Droegemeier, and D. Weber. Numerical Prediction of High-Impact Local Weather: A Driver for Petascale Computing. *Petascale Computing: Algorithms and Applications*, pages 103–124, 2007.

[248] Jae-Seung Yeom, Abhinav Bhatele, Keith Bisset, Eric Bohm, Abhishek Gupta, Laxmikant V. Kale, Madhav Marathe, Dimitrios S. Nikolopoulos, Martin Schulz, and Lukasz Wesolowski. Overcoming the scalability challenges of contagion simulations on Blue Waters. Technical Report 13-057, NDSSL, Virginia Bioinformatics Institute at Virginia Tech, 2013.

[249] Y. B. Zeldovich and R. A. Sunyaev. The Interaction of Matter and Radiation in a Hot-Model Universe. *Astrophysics & Space Science*, 4:301–316, July 1969.

[250] Gongpu Zhao, Juan R. Perilla, Ernest L. Yufenyuy, Xin Meng, Bo Chen, Jiying Ning, Jinwoo Ahn, Angela M. Gronenborn, Klaus Schulten, Christopher Aiken, and Peijun Zhang. Mature HIV-1 capsid structure by cryo-electron microscopy and all-atom molecular dynamics. *Nature*, 497:643–646, 2013. doi:10.1038/nature12162.

[251] Gengbin Zheng. *Achieving High Performance on Extremely Large Parallel Machines: Performance Prediction and Load Balancing*. Ph.D. thesis, Department of Computer Science, University of Illinois at Urbana-Champaign, 2005.

[252] Gengbin Zheng, Abhinav Bhatele, Esteban Meneses, and Laxmikant V. Kale. Periodic Hierarchical Load Balancing for Large Supercomputers. *International Journal of High Performance Computing Applications (IJHPCA)*, March 2011.

[253] Gengbin Zheng, Orion Sky Lawlor, and Laxmikant V. Kalé. Multiple flows of control in migratable parallel programs. In *2006 International Conference on Parallel Processing Workshops (ICPPW'06)*, pages 435–444, Columbus, Ohio, August 2006. IEEE Computer Society.

[254] Gengbin Zheng, Xiang Ni, and L. V. Kale. A Scalable Double In-Memory Checkpoint and Restart Scheme towards Exascale. In *Proceedings of the 2nd Workshop on Fault-Tolerance for HPC at Extreme Scale (FTXS)*, Boston, USA, June 2012.

[255] Gengbin Zheng, Lixia Shi, and Laxmikant V. Kalé. FTC-Charm++: An In-Memory Checkpoint-Based Fault Tolerant Runtime for Charm++ and MPI. In *2004 IEEE International Conference on Cluster Computing*, pages 93–103, San Diego, CA, September 2004.

[256] Gengbin Zheng, Terry Wilmarth, Praveen Jagadishprasad, and Laxmikant V. Kalé. Simulation-based performance prediction for large parallel machines. In *International Journal of Parallel Programming*, volume 33, pages 183–207, 2005.

Index